INTRODUCTION to ORGANIC PHOTOCHEMISTRY

INTRODUCTION to ORGANIC PHOTOCHEMISTRY

J. D. Coyle
The Open University

JOHN WILEY & SONS
Chichester · New York · Brisbane · Toronto · Singapore

Library of Congress Cataloging in Publication Data:

Coyle, J.D. (John D.)
Introduction to organic photochemistry.

Includes bibliographies and index.
1. Photochemistry. 2. Chemistry, Organic. I. Title.
QD708.2.C69 1986 547.1'35 85–29593

ISBN 0 471 90974 2 (cloth)
ISBN 0 471 90975 0 (paper)

British Library Cataloguing in Publication Data:

Coyle, J.D.
Introduction to organic photochemistry.
1. Photochemistry. 2. Chemistry, Organic.
I. Title.
547.1'35 QD708.2

ISBN 0 471 90974 2 (cloth)
ISBN 0 471 90975 0 (paper)

Printed and bound in Great Britain

CONTENTS

1 WHAT IS ORGANIC PHOTOCHEMISTRY?

Distinctive features of photochemical reactions 2
Absorption of light by organic molecules 9
Static properties of excited states 15
Dynamic properties of excited states (intramolecular) 20
Dynamic properties of excited states (intermolecular) 26
Mechanisms of excited-state processes 30
Methods of preparative photochemistry 37

2 PHOTOCHEMISTRY OF ALKENES AND RELATED COMPOUNDS

Geometrical isomerization 42
Electrocyclic processes 46
Sigmatropic shifts 51
Di-π-methane reaction 54
Other photoisomerizations 56
Addition reactions 58
Cycloaddition reactions 61
Photo-oxidation 69
Alkynes 72

3 PHOTOCHEMISTRY OF AROMATIC COMPOUNDS

Substitution reactions 77
Ring isomerization 86

Addition reactions 90
Cycloaddition reactions 91
Cyclization reactions 97

4 PHOTOCHEMISTRY OF ORGANIC CARBONYL COMPOUNDS

Bond cleavage 107
Hydrogen abstraction 115
Intramolecular hydrogen abstraction 119
Cycloaddition to carbon–carbon multiple bonds 126
Rearrangement of cyclohexenones and cyclohexadienones 131
Thiocarbonyl compounds 136

5 PHOTOCHEMISTRY OF OTHER ORGANIC COMPOUNDS

Imines, iminium salts and nitriles 142
Azo-compounds 148
Diazo-compounds, diazonium salts and azides 151
Nitrites and nitro-compounds 155
Saturated oxygen and sulfur compounds 159
Halogen compounds 162
Photohalogenation and photonitrosation 166
Photo-oxidation of alkanes 168

INDEX 172

PREFACE

Almost the whole of the energy received by the Earth comes from the Sun in the form of electromagnetic radiation of ultraviolet, visible or infrared wavelengths. The visible radiation is responsible for initiating vital processes in living organisms, such as photosynthesis (which provides material for the food chain and maintains the level of oxygen in the atmosphere), vision (through which higher organisms gain information about their environment) and the control of plant growth and seasonal development. The Sun's ultraviolet radiation plays a part in maintaining a low concentration of ozone in the upper atmosphere, which in turn absorbs most of the more harmful short-wavelength radiation and prevents it from reaching the surface of the planet. In addition to these natural phenomena, artificial sources of ultraviolet light find a range of technological applications, including the production of microelectronic circuits, the formation of polymerized coatings on manufactured articles and the sterilization of water for drinking. In most of these phenomena or applications the chemical process begins when an organic compound absorbs visible or ultraviolet light and undergoes a *photochemical* reaction.

Interest in organic photochemistry has also developed from a different direction, as synthetic organic chemists have incorporated into their strategic planning of syntheses, reaction types that can best be accomplished photochemically, or in some instances can be carried out only by a photochemical method. A great many of these organic photochemical reactions have been discovered relatively recently.

The purpose of this book is to provide an introductory account of the major types of organic photochemical reaction, so that students and others with some prior knowledge of basic organic chemistry can appreciate the differences between processes that occur through an electronically excited state and those that occur directly from the electronic ground state. At the end of each chapter are a few suggestions for further reading, so that the interested reader can follow up the major aspects in greater detail. These references are mainly to recent review articles or chapters, which in turn will provide access to the primary literature.

The material is organized according to the functional type of the substrate, in a way similar to that adopted in many general textbooks on 'ordinary' (thermal) chemistry. Other classifications can be made, but this one seems best suited to our present purposes. Some classes of substrate, such as ketones or alkenes, are very important in their ground-state reactions and in their excited-state reactions. Some that feature prominently in thermal chemistry, such as haloalkanes or carboxylic acids and their esters, are less important in a broad coverage of photochemistry; the reverse is also true, for instance with halo-substituted aromatic compounds. The emphasis in this book is on potentially useful organic photoreactions, with a bias towards usefulness in synthesis. For this reason there is little material on vapour-phase or short-wavelength photofragmentations, which have provided much of our detailed understanding of the way in which electronic excitation energy is channelled in a chemical transformation. However, a general mechanistic understanding is important if chemical knowledge is to be applied usefully, and this is provided within a wider framework of photophysical processes in the first chapter.

John Coyle

Milton Keynes
October 1985

CHAPTER 1

What is organic photochemistry?

A general working definition is that photochemistry is 'chemical change brought about by light'. Usually 'light' is taken to mean electromagnetic radiation in the visible and ultraviolet range (approximate wavelengths 700 to 100 nm), and 'chemical change' is interpreted broadly to include most events that occur at the molecular level after absorption of a photon, even if they do not lead to *overall* chemical change. A distinction can be made between photophysical processes, which lead through changes in energy and electronic structure to a regeneration of the same molecular species that existed prior to absorption, and photochemical changes, which produce new molecular entities. The distinction is clearly important as far as synthetic applications are concerned, but it is neither convenient nor helpful to dissociate the two types of process completely, because an understanding of the processes that bring about chemical change (and hence the ability to manipulate such changes) requires an understanding of the associated, and often competing, physical processes.

As far as this book is concerned the emphasis is on the photochemistry of organic compounds that produces overall chemical change. There is a bias towards reactions that are of potential synthetic value, reflected for instance in the choice of examples to demonstrate that reasonable isolated yields of products can be achieved. Some reactions that are of more theoretical interest are omitted or given lesser prominence, such as the fragmentation reactions of saturated compounds (for example, alkanes or haloalkanes) that occur when very short wavelength (very high energy) light is absorbed. Large areas of interesting and important photochemistry lie largely outside the scope of the book, either because they involve inorganic rather than organic species (for example, photochemical reactions occurring in the Earth's atmosphere, or photographic processes based on silver halides), or because the importance of the topic lies as much in complex subsequent reactions as in the initial step following absorption of a photon (for example, photosynthesis, vision, medical phototherapy or photopolymerization).

The purpose of this first chapter is to provide the wider photochemical context, and the terms and definitions, that make for a readier comprehension of the photoprocesses described in later chapters. In it are provided some of the general features of photochemical reactions that distinguish them from reactions that occur without light (thermal reactions), relevant features of the absorption of electromagnetic radiation by organic molecules, a description of the commonly encountered electronic transitions and electronically excited states that are part of organic photochemical processes, static properties of electronically excited states, dynamic properties of excited states related to intramolecular change or to interaction with a second molecular species, general comments on photochemical reaction mechanisms and their kinetic features, and finally a brief review of practical aspects of carrying out photochemical reactions in the laboratory for preparative purposes.

Distinctive features of photochemical reactions

Absorption of a photon by an organic molecule provides what is called an electronically excited state, which is the starting point for subsequent reaction steps. This is a molecular species whose electrons are arranged in a way that is *not* the lowest energy configuration, as seen in a stylized molecular orbital diagram (Figure 1.1) showing an array of orbitals on a vertical energy scale. We will look later in this section at the particular types of orbital commonly used to describe organic compounds. In the ground state (a) the available electrons occupy, in pairs, the lowest-energy orbitals, and the overall energy of the molecule is the lowest possible. In an electronically excited state (b) one electron occupies a higher-energy orbital, rather than the lowest-energy orbital available to it, and the overall energy of the molecule is higher than in the ground state. The electronically excited state is a distinct species, with a finite lifetime and with physical and chemical properties that differ from those of the ground state.

Electronically excited states can be produced by a variety of

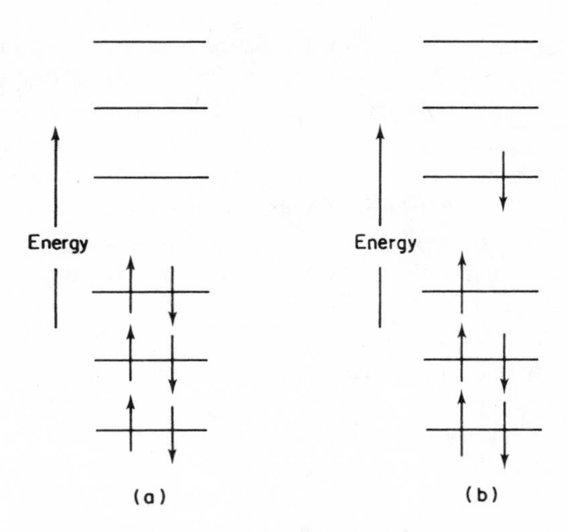

Figure 1.1 Stylized molecular orbital diagram for an organic molecule, showing some of the orbitals and their occupancy (a) in the ground state, and (b) in an electronically excited state.

methods, for example in an electric discharge, in a flame, by the use of ionizing radiation, or by chemical reaction. The last of these methods often leads to chemiluminescence, the emission of visible light from a chemical reaction, since light emission is a common property of electronically excited states. However, by far the most convenient and commonly used method for producing excited states is by absorption of a photon of visible or ultraviolet light. Electromagnetic radiation of these wavelengths has the energy required to convert a ground state into one of the lower-lying excited states, that is one in which an electron is promoted from a relatively high-energy occupied orbital to an orbital, unoccupied in the ground state, that is of relatively low energy. Typical numerical values for the energies are given later in the section.

We can digress at this point to consider two 'laws' of photochemistry that feature in most introductions to the subject. First, the Grotthuss–Draper law states, in essence, that only light absorbed by a molecule can be effective in bringing about chemical change. This may seem obvious when considered at the molecular level—if the photon energy is not made available to the molecule by absorption, an electronically excited state cannot be produced, and no photochemical change can result. The law's importance is in its practical application: it is not unknown for a photochemical reaction to be attempted, the substrate to be recovered unchanged, and only then for it to be realized that the light source or glassware used was such that the light reaching the sample was of wavelengths outside the range of the substrate's absorption spectrum!

The second law is the Stark–Einstein law, whose re-statement in current terminology is that the primary photochemical act involves absorption of just one photon by a molecule. This holds true for the vast majority of processes; exceptions to it arise largely when very intense light sources, such as lasers, are employed, and the probability of concurrent or subsequent absorption of two or more photons is no longer negligible.

Photochemistry, then, is concerned with the reactions of electronically excited states, and there are some general differences between these states and the corresponding ground states that greatly affect the nature of photochemical reactions. First, an electronically excited state is considerably more energetic than the ground state, which means that on thermodynamic grounds a wider range of reactions is open to it. Consider the situation depicted in Figure 1.2, which is a standard free-energy diagram showing the ground state of a molecule, an excited state of the same molecule, and a potential product of chemical reaction.

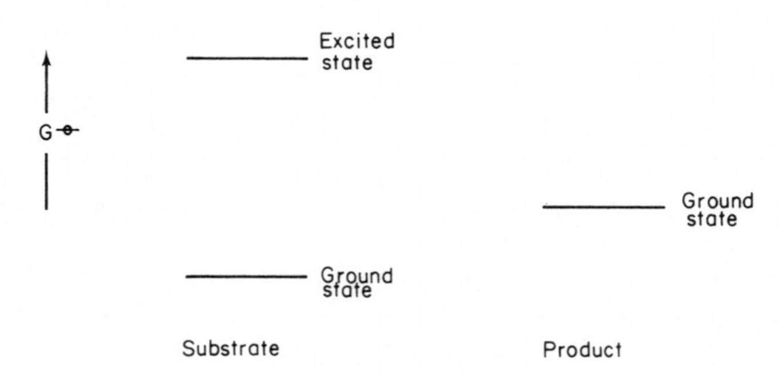

Figure 1.2 A standard free-energy diagram.

The reaction leading from ground state to product (in their standard states) involves an increase in free energy, and this will not occur spontaneously. If the free-energy difference is not large, a small equilibrium concentration of product might be achieved in homogeneous solution; for an intermolecular reaction, an increased equilibrium concentration of product may be obtained by using excess of the second reagent. However, the photochemical reaction that leads from excited state to product involves a decrease in free energy and is much more likely to occur spontaneously. This offers a rationalization of the fact that many 'energy-rich' compounds, such as those with high ring-strain, can be made photochemically. It should be borne in mind that thermodynamic feasibility is not a sufficient condition for a reaction to be observed—there may be kinetic barriers—but it is a normal prerequisite, and so the potential range of products available by photochemical reaction is much wider than the range associated with thermal processes.

A second major difference in the excited state is the very different electron distribution from that in the ground state. This has a considerable influence on the chemical changes observed; after all, organic reaction mechanisms are generally rationalized or proposed on the basis of electron distributions. For example, the susceptibility of ketones to attack by nucleophiles at the carbonyl carbon is attributed (1.1) to the polarization of electrons in the C=O bonds. The (n,π^*) excited state of a ketone (the terminology is explained later in this section) does not have such a large polarization; its outstanding feature, as far as electron distribution is concerned, is the odd-electron character of the oxygen atom, and much of its chemistry can be successfully rationalized in terms of processes derived from an oxygen-centred radical-like species. The general outcome of such changes in electron distribution on excitation is that expectations based on ground-state analogies are often overturned in photochemical reactions: many aromatic compounds, for instance, undergo photochemical substitution by nucleophiles (not electrophiles, as in the ground state), or photochemical addition to give products in which the aromatic ring is no longer present.

$$Nu^- \longrightarrow R_2C^{\delta+}{=}O^{\delta-} \qquad (1.1)$$

The symmetry properties of orbitals, and the pattern of electron occupation of the orbitals, is important in concerted reactions, that is, those that transform substrates directly into products without going through intermediates. A large group of concerted reactions called pericyclic reactions have been widely studied for alkenes and related compounds, and the Woodward–Hoffmann rules were formulated to codify the influence of orbital properties on the relative ease of alternative stereochemical courses for these reactions. Theoretical bases for understanding pericyclic reactions are more appropriately described in a general chemistry textbook. The relevant outcome in the photoreactions of alkenes and related compounds (see Chapter 2) is either that the observed stereochemical course for a photochemical reaction is normally the opposite of that observed for analogous thermal reactions, or that a reaction that occurs readily in the ground state does not occur readily in the excited state (and vice-versa). Once again the different electronic properties of an excited state are seen to have a major influence on the chemistry that the state undergoes.

A further difference between the electronic properties of ground states and excited states whose importance is increasingly recognized lies in the electron-donating or electron-accepting abilities of

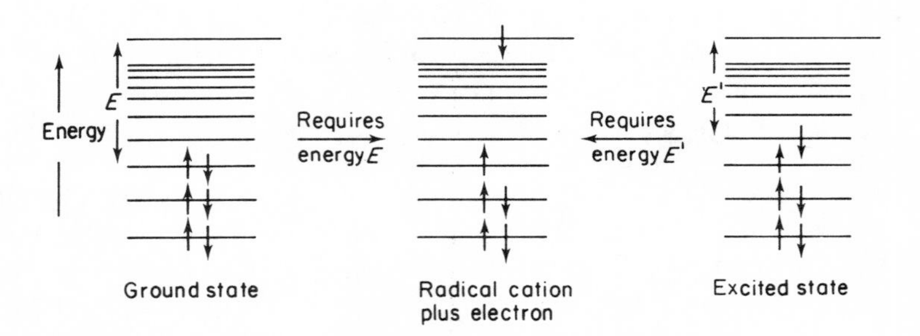

Figure 1.3 Diagrammatic rationalization of the better electron-donor properties of an excited state.

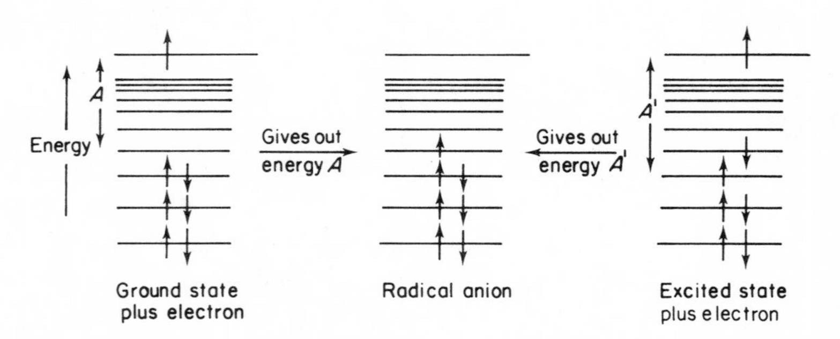

Figure 1.4 Diagrammatic rationalization of the better electron-acceptor properties of an excited state.

the states. An electronically excited state is both a better donor *and* a better acceptor of electrons than is the corresponding ground state. This can be rationalized in qualitative terms by reference to orbital energy level diagrams, assuming that the orbital energies are not greatly affected by removal of an electron from, or addition of an electron to, the molecule. A measure of the electron-donating ability of a species is its ionization potential, the minimum energy required to remove an electron completely from the species. This energy can be equated approximately to the difference in energy between the highest occupied molecular orbital and the ionization limit, which corresponds to complete removal of the electron and formation of a radical cation. This energy is considerably lower for the excited state (Figure 1.3), in which one electron is clearly closer to the ionization limit than for the ground state. The energy released when an electron comes from outside the influence of the nuclei of a species to occupy the lowest-energy available orbital is called the electron affinity, and it offers a measure of electron-accepting ability. By an analogous argument this energy is greater for the excited state than for the ground state (Figure 1.4), because in the former a lower-energy orbital is only half-filled.

Because of this difference in electron-donor and electron-acceptor properties, excited states have very different redox properties from those of related ground states. The effect is so marked that many photochemical processes begin with a complete transfer of an electron from (or to) an excited state (1.2), and the subsequent chemistry is that of radical cations and radical anions, species that are regarded as unusual in ground-state organic reactions. The importance of photochemical electron transfer is underlined by its extensive involvement in photobiological processes such as photosynthesis.

$$\mathrm{A} \xrightarrow{h\nu} \mathrm{A}^* \xrightarrow{\mathrm{B}} \mathrm{A}^{\cdot+} + \mathrm{B}^{\cdot-} \quad \text{or} \quad \mathrm{A}^{\cdot-} + \mathrm{B}^{\cdot+} \qquad (1.2)$$

In addition to these differences between excited-state and ground-state properties that influence chemical behaviour, there are some practical considerations that give photochemistry its distinctive features. In a thermal reaction, heat energy is normally supplied in an indiscriminate way to all the species in the reaction mixture—substrates, solvent and products—and this makes it difficult, for example, to prepare heat-sensitive compounds. In a photochemical reaction light can, in principle, be supplied selectively to just one species in the mixture. This is done by choosing a suitable exciting wavelength on the basis of the absorption characteristics of all the species present, or a suitable sensitizer on the basis of excited-state energies. In this way a thermally labile product may be generated at ambient temperature, or even more labile compounds by carrying out the irradiation at low temperature. The temperature at which photochemical reactions can be carried out can be lowered indefinitely, and irradiations at near to absolute zero have been used to prepare extremely reactive species, such as benzyne, under conditions where their lifetime is long enough for spectroscopic properties to be measured. Although this selective nature of photochemical reactions has limitations, for instance those imposed by the limited accessibility of some (monochromatic) wavelengths of ultraviolet radiation, it can be a great advantage in planning the synthesis of sensitive compounds.

Many thermal reactions are effectively irreversible under the conditions employed, but some are reversible and an equilibrium position is reached between substrates and products. The position of equilibrium depends on the standard free energy difference between the two ($\Delta G^{\ominus} = -RT \ln K$) and on reagent concentrations, and it varies with temperature. Such considerations rarely apply to photochemical reactions, the overwhelming majority of which are effectively irreversible (1.3), and the products are not in thermodynamic equilibrium with the reacting excited state.

$$\begin{array}{l} A \xrightarrow{h\nu} A^* \rightarrow B \\ \qquad\quad B \nrightarrow A^* \end{array} \tag{1.3}$$

$$\begin{array}{l} A \xrightarrow{h\nu} B \\ B \xrightarrow{h\nu} A^* \end{array} \tag{1.4}$$

A distinct situation is when a photochemical reaction has a corresponding reverse reaction (1.4). However, these are different reactions, because the first reaction goes through an excited state of A, whereas the second reaction goes through an excited state of B. This can lead to a practical reversibility, but it should not be called an equilibrium—the term 'photoequilibrium' is sometimes used, but it can be misleading. If such a situation is reached where the rate of conversion of A to B equals the rate of conversion of B to A, the system is said to have reached a photostationary state. The factors governing the relative composition of this steady state are the absorption coefficients of A and B at the wavelength used in the experiment, and the relative efficiencies of the forward and reverse photochemical reactions. The main controllable variable is the wavelength of irradiation; the concentration of reagents can also influence the composition if intermolecular reactions are involved. Specific examples arise particularly in the photochemistry of alkenes (Chapter 2).

Throughout this section on the differences between photochemical and thermal reactions, mention has been made of the electronically excited states that are key species in photoprocesses. We need now to look in more detail at the production of such excited states by absorption of light, and at the nature of the excited states of organic molecules.

Absorption of light by organic molecules

In the context of an introduction to organic aspects of photochemistry, a simple molecular orbital description of the electronic structure of organic molecules provides the most convenient qualitative framework in which to discuss the formation of electronically excited states by the absorption of light. It is normally assumed that the inner-shell electrons of the constituent atoms of a molecule remain unaltered in the molecule itself; linear combinations of the remaining, valence-shell atomic orbitals then provide molecular orbitals that can be used to describe the 'outer' electronic structure in the molecule.

If single atomic orbitals on each of two adjacent atoms are combined, they produce two molecular orbitals, one of higher energy and one of lower energy than the separate atomic orbitals (Figure 1.5). The lower-energy orbital is called a bonding orbital, and in a typical situation where there is a two-electron bond between the atoms, a pair of electrons will occupy the bonding orbital in the ground-state electronic configuration. The higher-energy molecular

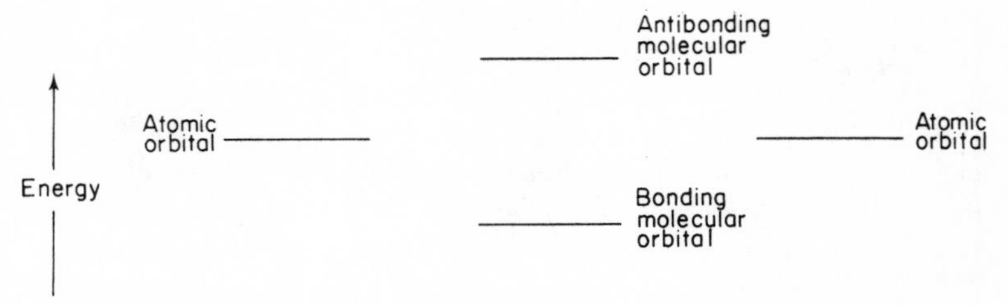

Figure 1.5 Effect on orbital energy of a linear combination of atomic orbitals.

orbital is called an antibonding orbital; it is unoccupied in the ground state, but it may be occupied by one electron in an electronically excited state of the molecule.

Orbitals that are completely symmetrical about the internuclear axis (such as those formed by combination of atomic *s* orbitals) are called sigma (σ) or sigma-star (σ^*) orbitals, according to whether they are bonding or antibonding, respectively. Orbitals that are antisymmetric about a plane that includes the internuclear axis (for example, those formed by combination of parallel atomic *p* orbitals) are called pi (π) or pi-star (π^*) orbitals. In principle orbitals can be constructed (mathematically) that cover all the nuclei of a particular model. In practice two simplifications can be made that afford a more readily visualized model. First, it is assumed that there is no interaction between σ and π orbitals because of their different symmetry properties, and they can be considered separately. Secondly, it is assumed that the σ-framework of an organic molecule can be described in terms of *localized* orbitals, each covering two nuclei only. *Delocalized* orbitals that cover more than two nuclei appear in this simple model only for π-bonding in conjugated organic molecular systems. A third type of orbital used in such a qualitative description is denoted as an *n*-orbital. These orbitals are usually nonbonding (that is, they are of much the same energy as in the corresponding isolated atom), and a pair of electrons occupying an *n*-orbital is often envisaged as a 'lone pair' of electrons on a particular atom.

It needs to be stressed that this simple molecular orbital picture is not appropriate for all purposes, but it is convenient for visualizing the changes brought about by light absorption in organic molecules, and as a qualitative basis for describing the mechanisms of organic photochemical reactions.

Visible light and ultraviolet radiation are examples of electromagnetic radiation, which can be described in terms of an oscillating electric field and an oscillating magnetic field in planes that are perpendicular to each other and to the direction of propagation. This wave description of light is complemented by a particle description in which radiation is considered to be emitted, transmitted and absorbed in discrete units (photons), whose energy (E) is directly proportional to the frequency (ν) of the electromagnetic oscillation (1.5); the proportionality constant is Planck's constant (h).

$$E = h\nu \qquad (1.5)$$

Commonly used units are reciprocal second (s^{-1} or Hz) for frequency; nanometre (nm, 1 nm = 10^{-9} m) for wavelength ($\lambda = c/\nu$, where c is the speed of light); and joule (J) for energy. Because energy is frequently dealt with on a mole basis, the unit encountered is kJ mol^{-1}. Table 1.1 shows the relationship between these units for selected numerical values in the ultraviolet/visible/near-infrared range.

Table 1.1 Equivalence of frequency, wavelength and energy units for electromagnetic radiation.

	Frequency/s^{-1}	Wavelength/nm	Energy/kJ mol^{-1}
Far ultraviolet	3.00×10^{15}	100	1200
Ultraviolet	1.20×10^{15}	250	480
Ultraviolet	0.86×10^{15}	350	343
Blue-green	0.60×10^{15}	500	240
Red	0.43×10^{15}	700	171
Near infrared	0.30×10^{15}	1000	120

Other units that you may encounter, especially in connection with spectroscopic data are: for wavelength, Ångstrom (Å; 1 Å = 0.1 nm); for energy, electron volt (eV; 1 eV = 96.5 kJ mol^{-1}); for wavenumber ($\bar{\nu} = \lambda^{-1}$), reciprocal centimetre (cm^{-1}). Sometimes cm^{-1} is employed as though it were an energy unit, but this is only a convenient shorthand. Some organic photochemists still use calorie (cal; 1 cal = 4.18 J) as their preferred energy unit.

When a photon interacts with a molecule, the mutual perturbation of electric fields may be of no lasting consequence. Alternatively, the photon may cease to exist, its energy is transferred to the molecule, and the electronic structure of the molecule changes; the photon has been absorbed and the molecule excited. Absorption occurs in an extremely short period of time ($\sim 10^{-15}$ s), and it can be assumed that the positions of the nuclei in the molecule do not change during this period—the major change is in the electronic structure. This electronic change can be pictured on the basis of our simple molecular orbital model in terms of a change in the pattern of occupation in a fixed set of orbitals, as already shown in Figure 1.1. The assumption that the set of orbitals is the same for both excited and ground states (called the one-electron excitation approximation) is valid for most absorption processes, although sometimes it is necessary to use orbitals not normally employed for a description of the ground state.

To take a more particular example, Figure 1.6 depicts the orbitals of the carbon–carbon bonds of an alkene, and it shows (a) the ground-state configuration, and (b) an excited-state configuration that might be obtained by absorption of a photon of wavelength around 180 nm. The excited state is called a pi, pi-star (π, π^*) excited state, and the electronic transition leading to it is a pi to pi-star ($\pi \rightarrow \pi^*$) transition.

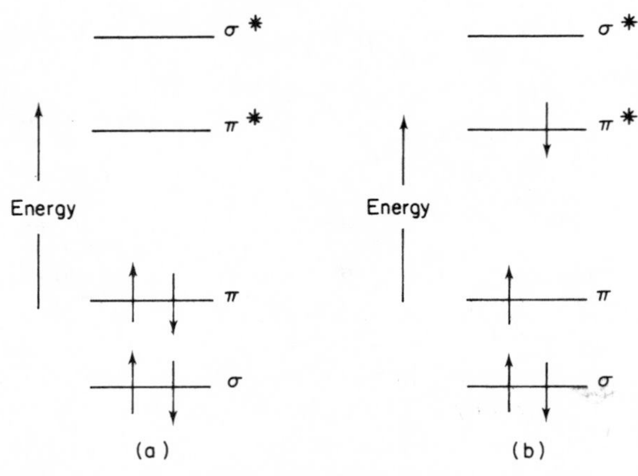

Figure 1.6 Electronic configuration of an alkene, showing the occupancy of the carbon–carbon bond orbitals (a) in the ground state, and (b) in the (π,π^*) excited singlet state.

The highest-energy orbitals for molecules containing heteroatoms (such as nitrogen, oxygen or halogen) are usually non-bonding orbitals, and the lowest-energy transitions for such molecules are $n \rightarrow \pi^*$ (if multiple bonds are present) or $n \rightarrow \sigma^*$ (for saturated molecules). Although other types of excited state do need to be considered at times, the great majority of organic photochemical reactions carried out with radiation of wavelength greater than about 200 nm (a limit imposed in part by practical aspects of conducting preparative reactions in solution) can be accounted for in terms of three general classes of electronically excited state—(n,π^*), (π,π^*) and (n, σ^*). The only exceptional type of excited state that we shall meet briefly in this book is a Rydberg state, which can best be pictured as having an electron promoted to an orbital much larger than the 'core' of (two) nuclei and inner-shell electrons; the state thus bears some resemblance to an atomic situation with a positively charged centre and an outer region of electron density.

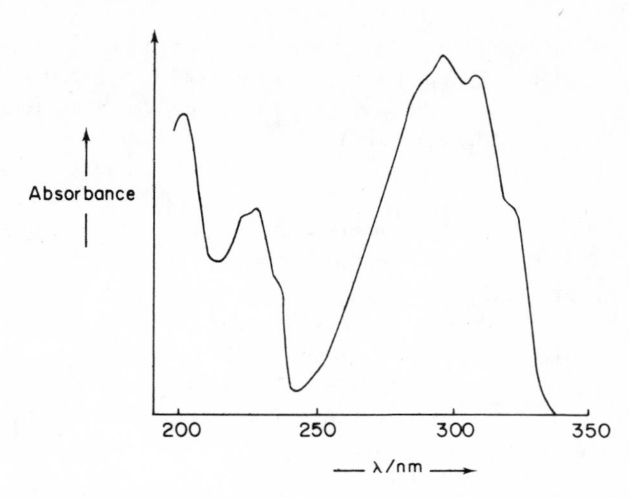

Figure 1.7 Electronic absorption spectrum of *trans*-stilbene.

The probability of absorption of a photon by a molecule, and its variation with wavelength, is reflected in the electronic absorption spectrum of the compound concerned. Figure 1.7 shows a typical spectrum for an organic compound in solution. At a particular wavelength the bulk absorption properties can be represented by the Beer–Lambert law (1.6); absorbance (D), which equals the base-10 logarithm of the ratio of incident light intensity (I_0) to transmitted light intensity (I), is directly proportional to the concentration (c) of the compound and the pathlength (l) of the radiation through the sample. The proportionality constant (ϵ) is called the absorption coefficient, or more specifically the molar absorption coefficient (units: $l\ mol^{-1}\ cm^{-1}$) if concentration is in $mol\ l^{-1}$ and pathlength in cm. The Beer–Lambert law is an empirical one, though it conforms to chemical 'common sense', and it holds for most conditions except when very high intensities of radiation are used (e.g. from lasers).

$$\log (I_0/I) = D = \epsilon c l \qquad (1.6)$$

Most solution-phase spectra of organic compounds show broad absorption bands, as in Figure 1.7, unlike atomic spectra, which consist of sharp lines. The main reason for this is that there are a large number of vibrational and rotational energy levels associated with polyatomic molecules, and absorption of a photon can result in conversion of a portion of its energy into vibrational or rotational energy. Hence for absorption to occur the energy of the photon does not need to match precisely the energy required simply to change the electronic configuration of the molecule. So there are a large number of slightly separated lines in the spectrum, and any fine structure that might remain is lost because solvent molecules cause a broadening of the lines as a result of their influence on the energy levels of the molecules. Despite this, the absorption spectrum can provide useful information about the electronic transitions that occur for a particular molecule, and about their relative energies. This is aided by a simple rule-of-thumb that bands corresponding to $\pi \rightarrow \pi^*$ transitions are usually more intense (ϵ 5000–100,000 l mol^{-1} cm^{-1}) than those corresponding to $n \rightarrow \sigma^*$ transitions (ϵ 100–1000 l mol^{-1} cm^{-1}) or $n \rightarrow \pi^*$ transitions (ϵ 1–400 l mol^{-1} cm^{-1}), combined with knowledge of the compound's molecular structure and a likely ordering of its orbital energies.

An electronically excited state has two unpaired electrons in different orbitals. In the states depicted so far these have been shown as having opposed spin, so that the state has overall zero spin and is a singlet state. This is the usual situation for excited states that are produced directly by absorption of a photon. If the two spins are parallel the state has a non-zero overall spin and is a triplet state. The state corresponding to the arrangement of electrons shown in Figure 1.8 is a (π,π^*) triplet state of an alkene, in contrast to the (π,π^*) singlet state depicted in Figure 1.6(b). The singlet and triplet states are said to be of different spin multiplicity, and they are distinct species, with differing properties and chemical reactions. The triplet state is of

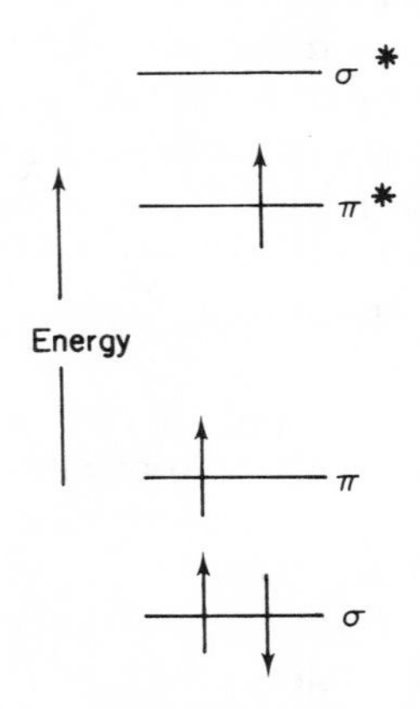

Figure 1.8 Electronic configuration of the (π,π^*) triplet state of an alkene.

lower energy (see the next section) because of the repulsive nature of interactions between electrons of the same spin, but it is not readily produced by direct absorption of a photon, because for most molecules the probability of a spin inversion occurring at the same time as absorption is extremely low. Triplet states are important in organic photochemistry, but they are formed indirectly.

Static properties of excited states

An excited state has a finite lifetime and so it has static properties, such as molecular shape (median bond lengths and angles) and dipole moment, like those of a ground-state molecule, that can in principle be determined experimentally. However, the lifetime of an excited state is short, often very short, and this restricts the range of techniques that can be employed to study such properties. Most of the available information comes from high-resolution absorption or emission spectra, particularly of small or symmetrical model compounds. The geometry of most other excited organic molecules has to be inferred from such results.

We have already seen that the process of light absorption is extremely fast, and at the instant after absorption the molecular geometry is unchanged and the nuclei occupy the same relative positions as in the ground state. This is an expression of the Franck–Condon principle and is useful in rationalizing the relative intensities within absorption or emission spectra. However, the excited state so produced does not have the most stable geometry for its particular electron distribution, and normally relaxation occurs by loss of vibrational and rotational energy to give the excited state in its 'equilibrium' geometry. The most stable molecular shape in the excited state may be quite different from that in the ground state, and this can have chemical consequences. For example, ethylene is a planar molecule in its ground state, but in the relaxed (π,π^*) excited state twisting has occurred around the erstwhile double bond (Figure 1.9a). Similarly, the (π,π^*) state of acetylene has a non-linear geometry (Figure 1.9b), and the (n,π^*) state of formaldehyde is non-planar (Figure 1.9c). The loss of stereochemical 'memory' in the relaxed (π,π^*) excited state of alkenes provides a rationalization of the photochemical *trans–cis* isomerization that is a characteristic reaction of alkenes (see Chapter 2).

The dipole moment of an excited state can be estimated from an analysis of solvent effects on absorption and emission spectra; the

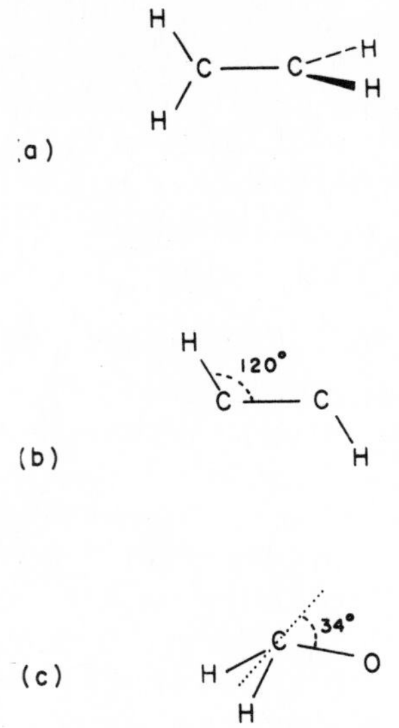

(a)

(b)

(c)

Figure 1.9 Geometry of (a) the lowest (π,π*) excited singlet state of ethylene, (b) the lowest (π,π*) excited singlet state of acetylene, and (c) the lowest (*n*, π*) state of formaldehyde.

wavelength of maximum emission is particularly sensitive to the nature and polarity of the solvent. This dipole moment, like that of a ground-state species, gives a measure of the overall electron distribution and may be a useful guide in predicting or rationalizing chemical behaviour. The dipole moment of formaldehyde, for example, is 2.3 D in the ground state, but only 1.6 D in the (*n*,π*) excited state, which points to a reduced polarization of the C=O bond in the excited state. Of particular interest are systems which have electron-donor and electron-acceptor groups linked through a conjugated system; these are quite polar in the ground state, but in the excited state they may have a very high dipole moment, indicating a high degree of charge transfer. The dipole moment of 4-nitroaniline (1.7) increases from 6 to 14 D on excitation; that of 4-amino-4′-nitrobiphenyl (1.8) from 6 to about 20 D, and that of 4-dimethylamino-4′-nitrostilbene (1.9) from 7.6 to about 26 D.

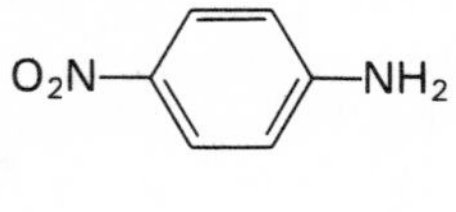

$O_2N-C_6H_4-NH_2$ (1.7)

$O_2N-C_6H_4-C_6H_4-NH_2$ (1.8)

$O_2N-C_6H_4-CH=CH-C_6H_4-NMe_2$ (1.9)

An excited-state property that is frequently used in the discussion of organic photochemistry is its energy. This is normally taken to be the energy difference between the relaxed excited state (in its most stable geometry and with no excess vibrational energy) and the ground state (also in its lowest vibrational level). A value for the energy of a lowest excited singlet state can be estimated from the absorption spectrum. If vibrational fine structure is apparent, as in the spectrum of Figure 1.10a, the longest-wavelength band apparent in the spectrum is often the one that corresponds to the energy required, and the wavelength can be converted to its equivalent energy. Spectra for most organic compounds in solution show little or no vibrational structure, as in the spectrum of Figure 1.10(b), and in these cases a reasonable guess can be made. Note that that absorption *maximum* does not correspond to the excited state energy.

A complementary, and more reliable, approach to singlet state energies is available from fluorescence emission spectra (see next section), which usually show more detailed fine structure than the absorption spectra. The shortest-wavelength band in the fluorescence spectrum corresponds to the excited-state energy. Even if the emission spectrum shows little vibrational structure, the energy can be estimated from the wavelength region where absorption and emission spectra overlap. For triplet state energies, estimates from

Table 1.2 Energies of excited singlet and triplet states

Compound	Nature of excited state	Singlet energy/ $kJ\ mol^{-1}$	Triplet energy/ $kJ\ mol^{-1}$
Acetone (Me_2CO)	n,π^*	365	325
Benzophenone (Ph_2CO)	n,π^*	320	290
Benzene	π,π^*	460	350
Buta-1,3-diene	π,π^*	430	250
Stilbene (PhCH=CHPh)	π,π^*	395	210

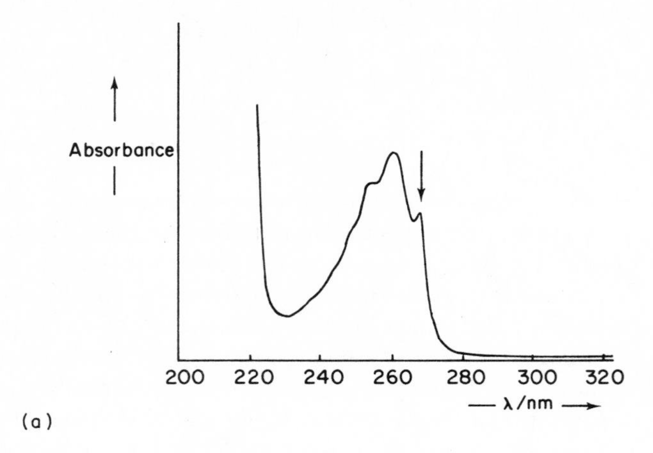

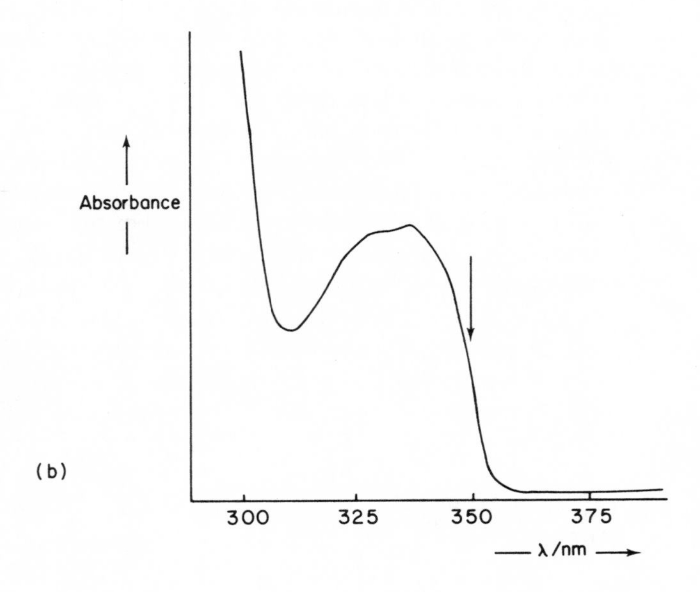

Figure 1.10 (a) Absorption spectrum of ethylbenzene, showing the lowest energy vibrational band; (b) absorption spectrum of acetylnaphthalene, showing an estimate of where the lowest energy vibrational band lies.

absorption spectra are rarely available, because singlet → triplet absorption spectra are very weak and difficult to obtain in useful form. Triplet energies are generally obtained from phosphorescence emission spectra or from energy-transfer studies (see the next two sections). A few typical excited-state energy values are given in Table 1.2.

Note that the triplet state energy is always lower than that of the singlet state. Also, although it may not be obvious from the small selection of data given, the difference in energy between singlet and triplet excited states is considerably smaller for (n,π^*) states than for (π,π^*) states. This is a diagnostic feature used in assigning the electronic nature of organic excited states. The energy of an excited state is a useful property in devising sensitized photochemical reactions, that is those in which the required excited state is produced indirectly by transfer of energy from another species rather than directly by light absorption. It can also be employed in thermodynamic calculations to consider the feasibility of particular transformations, and it is the appropriate energy value for this purpose, rather than the photon energy ($E = h\nu$), since much of the photon energy may be lost by vibrational and rotational relaxation to give the excited state from which reaction occurs.

An excited state has a characteristic lifetime under given conditions of temperature, solvent and concentration of substrate, and of other species in the solution. These lifetimes are short, varying from more than 10 s at liquid-nitrogen temperature for some of the longest-lived triplets to less than 10^{-12} s for some of the shortest-lived singlets. A small number of representative values are shown in Table 1.3.

Table 1.3 Lifetimes of excited singlet and triplet states

Compound	Singlet lifetime/s[a]	Triplet lifetime/s[b]
Acetone (Me_2CO)	2×10^{-9}	6×10^{-4}
Benzophenone (Ph_2CO)	5×10^{-12}	6×10^{-3}
Benzene	3×10^{-8}	6.3
Biphenyl (PhPh)	1.6×10^{-8}	4.6
Pyrene	4.5×10^{-7}	0.5

a In solution at room temperature.
b In a glass at 77 K.

The enormous range of lifetimes, more than thirteen orders of magnitude, encompasses the wide variety of photophysical and photochemical processes that can occur for an excited state, since lifetime is defined as the reciprocal of the sum of all the first-order (or pseudo-first-order) rate constants for processes that the excited state undergoes. This leads into the next two sections, which describe some of the dynamic, time-dependent properties of excited states that determine its lifetime.

Dynamic properties of excited states (intramolecular)

In the absence of interaction with another chemical species an electronically excited state can do one of two things—it can change into a different electronic state of the same compound, or it can change into a different compound. Conceptually the two are not very different. The first is a photophysical process, with which we are concerned in this section, and the second is a photochemical reaction, which is the subject of the remaining chapters of the book.

Intramolecular photophysical processes can be sub-divided into two groups—radiative (or luminescent) processes, in which a photon of ultraviolet or visible radiation is emitted, and non-radiative processes, in which no such emission takes place. Luminescence is a commonly observed property of electronically excited states, and it finds such practical applications as in optical whiteners for washing powders, security marking, and systems for automatic sorting of mail. The energy of the emitted light is less than that of the light used to excite the molecule originally, because some energy is converted into vibrational or rotational motion, and so visible light emission often occurs as a result of excitation by ultraviolet radiation.

When an excited singlet state emits a photon, the state is normally converted to the ground state, which is also a singlet state. Such a radiative process in which there is no overall change of spin is 'spin-allowed' and is called fluorescence. Most commonly it is the lowest excited singlet state (in its lowest vibrational level) whose emission is observed, regardless of which singlet state is formed initially by absorption. This observation is one aspect of what is known as Kasha's rule. The reason for it is that higher-energy singlet states generally undergo very rapid non-radiative decay to the lowest excited singlet state; emission from these upper states cannot compete effectively with such very fast decay, and in the exceptional cases where fluorescence from a higher singlet state is observed, its efficiency and intensity are low. Because emission is normally from the lowest excited singlet level, a fluorescence spectrum is not normally affected in its profile by the wavelength used to excite the sample. Another feature of fluorescence is that it is measured absolutely, unlike absorption which is measured by difference, and sensitive detectors allow very weak fluorescence to be observed, so that fluorescence can be used in chemical analysis for detecting, and measuring quantitatively, extremely small amounts of material.

Fluorescence spectra often exhibit clear vibrational fine structure, as shown in Figure 1.11, and the highest-energy (shortest-wavelength) emission band corresponds to the excited-state energy. The other vibrational bands are lower in energy because they derive from conversion of the lowest level of the excited state to higher vibrational levels of the ground state. It might be expected that the excited singlet energy so obtained from fluorescence information should be the same as the excited-state energy as determined from

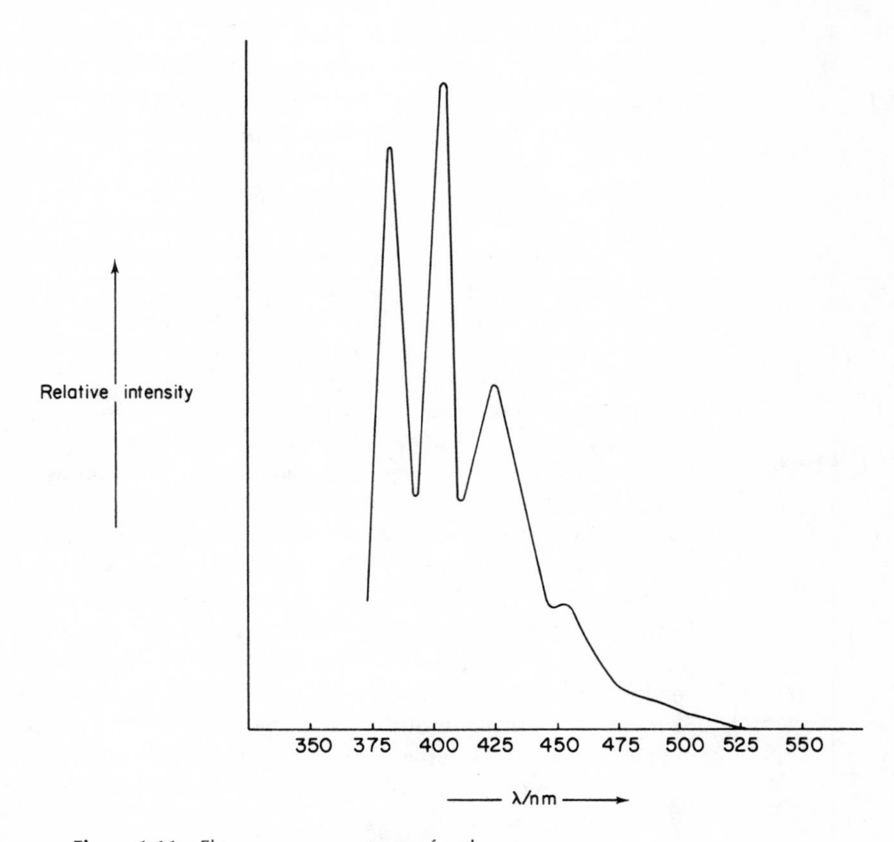

Figure 1.11 Fluorescence spectrum of anthracene.

absorption spectra. In fact the values are rarely identical, and the small difference arises because interaction with solvent molecules affects the electronic state energies differently.

Two useful fluorescence parameters are the quantum yield and the lifetime. Quantum yield is a property relevant to most photophysical and photochemical processes, and it is defined for fluorescence as in (1.10). More generally it is a measure of the efficiency with which absorbed radiation causes the molecule to undergo a specified change. So for a photochemical reaction it is the number of product molecules formed for each quantum of light absorbed:

$$\text{quantum yield for fluorescence } (\phi_f) = \frac{\text{number of photons emitted}}{\text{number of photons absorbed}} \tag{1.10}$$

Quantum yield values normally fall in the range 0 to 1.0, although in some specific situations a higher value for chemical reaction (though not for fluorescence) may be observed. The measurement of fluorescence quantum yields is not straightforward; even with unidirectional, monochromatic exciting light, fluorescent emission is in all directions and is polychromatic (which is relevant because detector response can vary with wavelength). A common practice is to measure quantum yields relative to that of a standard compound, excited under identical conditions, whose quantum yield is known from previous determination.

The lifetime of fluorescence is measured directly by following the decay of fluorescence intensity with time after the exciting radiation is interrupted. This may sound simple, but remember that singlet lifetimes are often in the nanosecond range! The techniques for measuring such short lifetimes are not important for the purposes of this brief review. Lifetime values are useful because they give, in reciprocal form, the overall sum of rate constants for singlet-state decay or reaction, which is one step towards estimating individual rate constants for excited-state processes. A further use is that the lifetime is sensitive to environmental conditions (for example, it varies from one solvent to another), and this has enabled fluorescence techniques to be employed to probe molecular environments in heterogeneous environments, such as at liquid–solid interfaces, in micelles or in biological tissue.

Phosphorescence is like fluorescence in many ways, except that in this type of radiative process the spin multiplicity of the species changes. Almost all phosphorescence for organic compounds involves luminescence that originates in the lowest excited triplet state, which decays to the ground state in the process. Because it is a 'spin-forbidden' process, phosphorescence generally has an associated lifetime that is considerably longer than fluorescence lifetimes for similar molecules. However, a direct comparison is not always meaningful because fluorescence and phosphorescence are usually studied under different conditions—fluorescence in fluid solution at room temperature, and phosphorescence in a rigid glass matrix at very low temperature, most commonly 77 K. Such conditions are used for phosphorescence studies because at room temperature in solution triplet states are generally deactivated very rapidly in bimolecular processes with other molecular species, which greatly reduces the lifetime and quantum yield of phosphorescence, and so makes it difficult to detect and study. There is no doubt that triplets are formed in fluid solution, because they can be detected by other methods such as flash photolysis, but their luminescence is very weak. Under the low-temperature conditions employed for phosphorescence studies, fluorescence may also be occurring (at shorter wavelengths, since excited singlet states have energies higher than the corresponding triplet states), and to differentiate between the two it is normal to incorporate a rapid shutter device into the instrument. This makes use of the difference in lifetime between fluorescence and phosphorescence, and enables the detector to be shielded from the luminescence during the very brief period after the exciting radiation is cut off, but it opens up subsequently to allow the longer-lived phosphorescence emission to reach the detector. In this way phosphorescence is studied without interference from fluorescence, and lifetimes and quantum yields can be determined.

Non-radiative (or radiationless) decay processes involve conversion of one electronic state into another without emission of light. Like radiative processes they can be divided into two categories according to whether or not there is an overall change in spin multiplicity during the process. If no spin-change occurs, the non-radiative process is called internal conversion. This is the means by which higher excited singlet states decay rapidly to the lowest excited singlet before further photophysical change or photochemical reaction occurs. Similarly, higher triplet states decay rapidly to the lowest triplet state by internal conversion. Internal conversion can also occur from the lowest singlet state to the ground (singlet) state in competition with fluorescence. For most molecules this particular internal conversion process is much slower than the processes by which higher singlet states decay. This arises because the rate of internal conversion is related inversely to the energy difference between the initial and final states (the smaller the energy gap, the faster the process), and the energy of the lowest excited singlet state (i.e. the energy difference between this state and the ground state) is usually much larger than the energy difference between the lowest singlet state and the singlet next higher in energy. Because internal conversion from the lowest excited singlet to the ground state is relatively slow, fluorescence can compete effectively with it, and for some molecules the quantum yield for this internal conversion process is negligibly small.

When non-radiative decay involves a change in spin multiplicity it is called intersystem crossing, and for organic molecules there are two important intersystem crossing processes. The first is the radiationless decay that competes with radiative (phosphorescent) decay of the lowest triplet state to the ground state. The second is the process that converts the lowest excited singlet state into the lowest triplet state and is therefore in competition with fluorescence and with internal conversion from the lowest singlet state. The importance of the second of these intersystem crossing processes is that it provides the means by which many triplet states are produced prior to phosphorescence or photochemical change, namely absorption of a photon to give the excited singlet state, followed by intersystem crossing to the triplet. As with other radiationless processes, the rate constant for intersystem crossing is related inversely to the energy gap between the initial and final states. This means that intersystem crossing to produce the triplet is fast, and efficient, for molecules which have a relatively small singlet–triplet energy difference, such as those ketones with lowest (n,π^*) states (see Table 1.2), but it is slow, and inefficient, for many molecules which have a much larger singlet–triplet energy gap, such as those alkenes with lowest (π,π^*) states. For this reason direct irradiation of aromatic ketones, for example, leads to chemical reaction from the lowest triplet state, whereas direct irradiation of conjugated dienes leads to reaction from the lowest singlet excited state.

After this brief description of the various intramolecular photophysical processes that can occur after excitation of a molecule, it is appropriate to bring them together by showing how they can be represented and related on a single diagram, known as a Jablonski

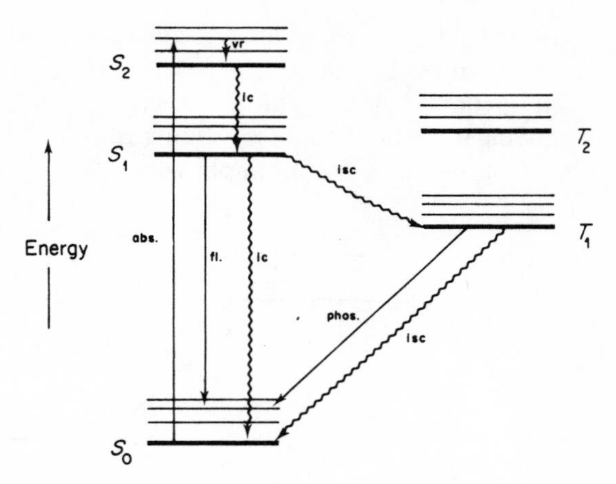

Figure 1.12 A Jablonski diagram.

diagram (Figure 1.12). In the left-hand portion of the diagram are the singlet states of the molecule, namely the ground state (S_0) and excited singlets (S_1, S_2, etc.) on a vertical scale of increasing energy; the subscripts 0, 1, 2, etc. refer simply to the order on the energy scale. In the right-hand portion are the triplet states (T_1, T_2, etc.) in order of increasing energy. This is a *state* diagram (compare the *orbital* diagrams of Figures 1.6 and 1.8), and each bold horizontal line represents a different electronic state of the molecule; that is, each represents a separate species. That the electronic states can have varying amounts of vibrational energy is indicated by a series of lighter horizontal lines ('vibrational levels') for each state. Photophysical processes are represented by lines connecting the states. Straight lines are used for radiative processes—vertically upwards for absorption, vertically downwards for fluorescence, and downwards at an angle for phosphorescence because it involves a change of multiplicity. Wavy lines represent radiationless transitions—vertically downwards for internal conversion, or for vibrational relaxation within a particular state, and at an angle for intersystem crossing. The degree of complexity in a Jablonski diagram is variable, according to the amount of detail required. In this figure only the first two excited singlet and triplet states are shown (there are more), only a few vibrational levels are indicated for each state, and radiationless processes are depicted as occurring directly to give the lowest vibrational level of the final state (they could be more correctly indicated as a horizontal line from initial to final state, followed by vibrational relaxation within the final state). This form is quite adequate for our purposes.

It might be helpful to look at a real example, drawn to scale and showing numerical values for some of the rate constants. Figure 1.13 is a Jablonski diagram for naphthalene. The data are obtained from

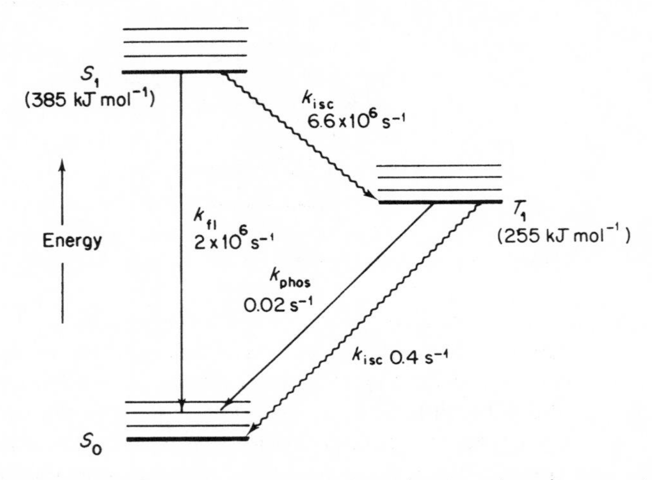

Figure 1.13 A Jablonski diagram for naphthalene, showing selected rate constants.

various sources, and measurements were taken under different conditions; precise comparisons may not be valid, but order-of-magnitude comparisons can be made. Naphthalene has lowest singlet and triplet states that are (π,π*) in character. Intersystem crossing from S_1 to T_1 and fluorescence from S_1 have comparable rate constants, and fluorescence is readily observable for this compound. Intersystem crossing from T_1 to S_0 is considerably faster than phosphorescence, and the efficiency of phosphorescence is not high (ϕ = 0.04).

Dynamic properties of excited states (intermolecular)

In addition to the intramolecular deactivation processes described in the previous section, there are intermolecular processes that can result from interaction with ground-state molecular species. If the result is permanent chemical change for both components, this is intermolecular photochemical reaction. Reversible chemical change is another possibility, with the production of metastable chemical species that revert, often rapidly, to the original molecules, both in their ground states. The metastable species may be the same intermediates that can also lead on to chemical products, in which case there is a partitioning between overall photochemical reaction and overall deactivation of the excited state. In the context of synthetic photochemistry the deactivation would be regarded as an energy-wasting pathway in competition with, and lowering the quantum yield of, the product-forming route. An especially important type of process that falls into this category is electron transfer, which has already been described as a distinctive feature of excited-state chemistry (1.2). Electron transfer in the reverse direction but with loss of the excitation energy can be an extremely fast reaction, and it represents a deactivation of the original excited state (1.11).

$$A^* + B \rightarrow A^{\cdot +} + B^{\cdot -} \rightarrow A + B \qquad (1.11)$$

There are, however, other ways in which excited-state decay can be accelerated by other species, which cannot be classified as reversible chemical reactions. Such processes can be represented generally by (1.12), where a star denotes electronic excitation. The excited state of A is said to be quenched by B. If B is converted into an electronically excited state (B*) during the process, an overall transfer of electronic energy takes place between the excited and unexcited partners of the interaction.

$$A^* + B \rightarrow A + B \text{ (or } B^*) \qquad (1.12)$$

In this general representation B might be a second, unexcited molecule of A, which would be the particular case of self-quenching. Self-quenching is a common phenomenon, one outcome of which is that the efficiency of luminescence in fluid solution may depend on the concentration of the substrate. For this reason fluorescence quantum yields are normally measured in very dilute solution (10^{-5} to 10^{-6} mol l^{-1}), when self-quenching is relatively unimportant. Of the several mechanisms for quenching, one involves the initial formation of a complex between A* and B. Such an *exciplex* (or *excimer* if B is a molecule of ground-state A) represents a metastable intermediate of an unusual kind: because it is electronically excited, and although the corresponding ground-state complex has no separate existence, the exciplex can luminesce. Exciplex fluorescence always occurs at lower energy (longer wavelength) than the normal fluorescence of

A*, and unambiguous evidence for exciplex formation is the appearance of a new, longer-wavelength fluorescence that increases in intensity as the concentration of B increases, with a corresponding decrease in the intensity of fluorescence from A*. Even when exciplex luminescence cannot be detected, such species may be postulated to account for other aspects of photophysical or photochemical processes.

Apart from exciplex formation, there are two major mechanisms by which energy-transfer quenching can occur. The first is known as the dipole–dipole (or Coulombic) mechanism, which operates through mutual repulsion of the electrons in the two molecules. It has the characteristic of being effective over relatively large distances, and in some systems is efficient at molecular separations up to 5 nm or more, which means that it does not require the molecules to move into close contact. A second mechanism is called the exchange mechanism, in which reorganization occurs within a transient complex formed on close approach of the molecules. This is a shorter-range phenomenon than dipole–dipole energy transfer.

The most commonly encountered types of energy transfer are those in which singlet energy is transferred from A to B (1.13), or those in which triplet energy is transferred (1.14). In most situations singlet energy transfer takes place mainly by the dipole–dipole mechanism, whereas triplet energy transfer occurs largely by the exchange mechanism. An over-riding consideration for both types is that efficient transfer occurs only if it is energetically favourable, that is if the excited state energy of A is greater than that of B.

$$\mathrm{A}^* \ (S_1) + \mathrm{B} \ (S_0) \rightarrow \mathrm{A} \ (S_0) + \mathrm{B}^* \ (S_1) \qquad (1.13)$$

$$\mathrm{A}^* \ (T_1) + \mathrm{B} \ (S_0) \rightarrow \mathrm{A} \ (S_0) + \mathrm{B}^* \ (T_1) \qquad (1.14)$$

Electronic energy transfer is an important phenomenon in photochemistry. For example, it plays a vital part in such processes as the light-harvesting steps in the mechanism of photosynthesis, and in other biological systems triplet quenchers such as β-carotene provide protection against photochemical degradation. In chemical studies quenching can be employed to obtain both qualitative and quantitative information. If a photochemical reaction forms different products from the singlet and triplet excited states, and if the products of the triplet reaction are unwanted, the singlet-derived products can be generated alone by carrying out the irradiation with an added triplet quencher in appropriate quantity. Conjugated dienes are widely used as triplet quenchers, such as penta-1,3-diene ($CH_3CH{=}CHCH{=}CH_2$), which is a low-boiling liquid. These are transparent to the longer wavelengths of ultraviolet radiation (and so do not themselves absorb the exciting light), they have a low triplet energy (and therefore are capable of quenching triplet states of many other compounds), and they have a high excited singlet energy (and therefore, for most substrates, do not at the same time cause singlet quenching by energy transfer). The relative energies for penta-1,3-diene, and those for acetone for comparison, are shown in Figure 1.14.

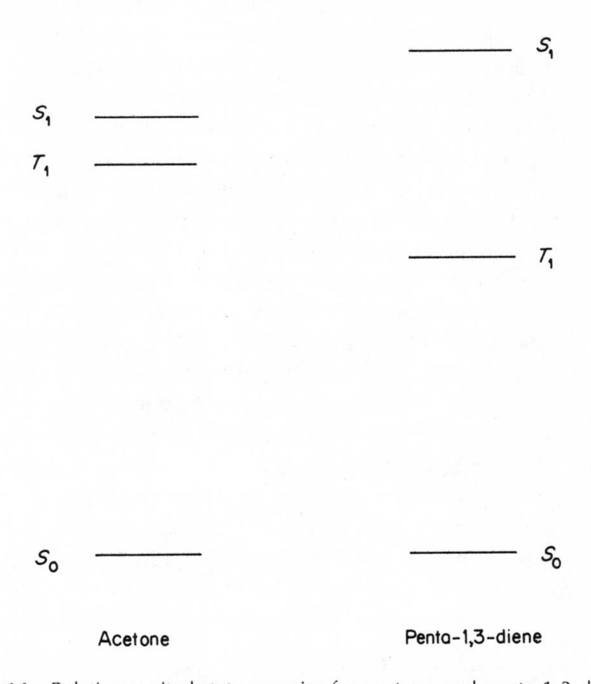

Figure 1.14 Relative excited state energies for acetone and penta-1,3-diene.

The rate of quenching an excited state is directly proportional to the concentration of quencher, and hence quantitative studies using a range of quencher concentrations can lead to information from which excited state lifetimes are derived. Use of a series of quenchers with different energies allows an estimate to be made of the excited state energy, especially for triplet states, since the efficiency of quenching falls markedly when the quencher triplet energy is lower than the energy of the reactive triplet state.

Energy transfer can be employed to produce an excited state of a substrate rather than to quench it, that is to generate and study B* in (1.13) or (1.14). This represents a good method for making triplet states (which cannot be readily obtained directly by $S_0 \rightarrow T_1$ absorption of light), especially those whose formation by intersystem crossing from the excited singlet state is inefficient. In the example given above of conjugated dienes as triplet quenchers, such triplet energy transfer is also the best method of forming the triplet states of dienes for subsequent reaction. When applied in this sense, the name sensitization is given to an energy transfer process. Compounds which themselves undergo efficient intersystem crossing can act as triplet sensitizers, the most common being aromatic ketones such as benzophenone (Ph_2CO). Sensitizers find important applications in colour photography, to provide a mechanism by which the available light (green, for example) can be used to promote photochemical change in a substrate (silver halide) that does not absorb it.

Oxygen occupies a special position in organic photochemistry. It is present in all reaction mixtures exposed to the air, it is a good electron-acceptor (forming the reactive $O_2^{\bullet -}$ radical anion), and it has, unusually, a triplet ground state and a very low-energy excited singlet state. The last property makes oxygen a very good triplet quencher, and excited singlet oxygen is very reactive towards organic compounds (see Chapter 2). Oxygen can therefore interfere in many preparative photoreactions and in quantitative studies of photophysical processes. It is normal practice to remove dissolved oxygen by bubbling an inert gas (such as nitrogen or argon) through the solution, or by operating a sequence of 'freeze–pump–thaw' cycles to liberate dissolved gases and pump them away.

Mechanisms of excited-state processes

Elucidation of the detailed molecular pathway by which a reaction (chemical or photochemical) occurs is of interest because it provides a framework to which different aspects of the reaction can be related. The synthetic chemist uses mechanistic information to rationalize or predict the effects of changes in controlled variables, such as solvent, temperature or concentrations of reagent, on the yields of products and on the optimum time for reaction. He can also use such information to assess the likely outcome of reactions that have not previously been attempted. Photochemical mechanisms have features that are additional to those investigated for thermal reactions, and this survey of excited-state mechanisms is concerned mainly with these distinctive characteristics.

The beginning and end points of a photochemical reaction pathway are the structures of the starting materials (substrates) and the isolated products. Elucidation of product structures can be carried out by conventional methods. Structure determination for products derived from labelled substrates, such as those with isotopic labels or with extra substituents, or from substrates with distinctive stereochemical features, can result in the elimination of certain mechanistic possibilities and provide support for others. Two key questions for photochemical mechanisms, as for thermal mechanisms, are whether or not a reactive intermediate (such as a biradical) lies on the reaction pathway, and if so, what are the rate constants for reaction steps subsequent to its formation. Questions that are peculiar to photochemical mechanisms may be expressed:

1. Which low-energy electronically excited states of the substrate might be on the reaction pathway, and which of these is formed initially (by light absorption or by energy transfer)?
2. Which excited state is the reactive one, in that it initiates the changes in chemical bonding?
3. What is the energy of the reactive excited state?
4. What is the lifetime of the reactive excited state, and the rate constant for the primary step leading from it to product (in a concerted mechanism) or to a reactive intermediate?

The major sources of information for answers to these questions are the absorption and luminescence spectra of the substrate, direct spectroscopic measurements of excited states and other reactive intermediates, and energy-transfer studies. In this section the intention is to outline the nature of evidence that can be obtained, and the sort of mechanistic conclusions that can be drawn, without providing detailed discussion of the methods involved.

In the earlier section of this chapter dealing with the absorption of light by organic molecules, we saw that absorption spectra can provide information about the low-lying singlet excited states of a molecule, and in particular about the electronic nature and the energy of the lowest excited singlet. The energy is estimated from the position of the longest-wavelength vibrational band in the spectrum, which may be clearly identifiable or may be part of a general envelope of absorption. The nature of the transition corresponding to a particular absorption band is indicated by the relative intensity of the band, which reflects the operation of 'selection rules' based on the spatial overlap and symmetry of the two orbitals concerned. For example, it is usually possible to distinguish readily between $n \rightarrow \pi^*$ and $\pi \rightarrow \pi^*$ absorption bands for a compound that is expected to give rise to both, because the former bands are quite weak compared with the latter. Additional information about the nature of the transition may come from an analysis of the change in its position (λ_{max}) in going from a non-polar to a polar solvent: absorption bands arising from an $n \rightarrow \pi^*$ transition show a pronounced blue shift, to shorter wavelength, with such a solvent change. The interpretation of this effect is not settled, but it is a useful diagnostic tool. In most instances it is the lowest excited singlet state that is of interest, since reaction from higher states is precluded by very rapid internal conversion from these states to the lowest state, and hence the reactive state is the same irrespective of which singlet is produced initially. There are a few exceptions to this generlization, and when different products are formed from a higher excited state there may be a pronounced effect of irradiation wavelength on product distribution, which can be related to the bands in the absorption spectrum.

Luminescence spectra provide information about singlet states (fluorescence) and triplet states (phosphorescence), and most often it is the lowest-energy excited singlet and triplet states that luminesce. As already discussed, the energy of the emitting state can be estimated from the shortest-wavelength vibrational band in the spectrum, and the lifetime may be measured by monitoring the decay of luminescence intensity. The electronic nature of the state cannot be assigned with certainty on the basis of luminescence properties alone, although there are some useful guidelines. For example, as can be seen in the short list of triplet lifetimes given in Table 1.3, the lifetimes for the two ketones at 77 K are two or three orders of magnitude shorter than those for the three aromatic hydrocarbons, and this is typical of the shorter triplet lifetime of (n,π^*) states compared with (π,π^*) states.

Although spectroscopic studies provide useful information about excited states produced by light absorption, or about excited states that luminesce, the question remains as to which (if any) of these states is the chemically reactive one. Spectra alone do not provide many indications about this, but energy-transfer studies are often helpful. For example, if a photochemical reaction is quenched by an added compound known to be a triplet quencher, it is most likely that the chemically active state is a triplet rather than a singlet; if the same quencher also reduces the intensity of phosphorescence in a similar way, the reactive triplet is almost certainly the same as the luminescent triplet. Similarly, if a photochemical reaction can be sensitized by a compound known to be a triplet sensitizer, giving the same products in the same ratios, a reactive triplet state is again implicated. Negative results from quenching studies need to be interpreted with caution, since they may arise because of particular features of the experiment design (e.g. choice of a quencher with too high a triplet energy) or from the operation of very fast primary reaction steps that make triplet quenching inefficient. In principle singlet sensitization and singlet quenching studies can be carried out, but in practice it is often not possible to select suitable compounds that meet the requirements for effective operation but are not also triplet sensitizers or quenchers. If a range of triplet quenchers or sensitizers is used and the efficiency of photochemical reaction monitored, it is possible to estimate the energy of the reactive triplet state, and this may be compared with the energy value measured for a phosphorescent state. All of these energy-transfer studies provide information about the identity of the excited state responsible for initiating chemical reaction.

The reactivity of an excited state is most appropriately quantified in terms of the rate constant for the first step on the reaction pathway leading to the eventual chemical product(s). It is not easy to obtain a value for this directly, and indirect methods are usually employed. The quantum yield for fluorescence has already been defined (1.10), and the quantum yield for product formation is derived from measurements of the amount of product formed in a given time and the amount of product formed under exactly the same conditions in a photochemical reaction whose quantum yield is already known (a chemical actinometer). This type of quantum yield measurement is readily carried out in a 'carousel' apparatus which ensures that two (or more) reaction tubes rotating with respect to a light source receive an equal intensity of light. Careful design of the experiment and choice of appropriate actinometer are necessary for reliable results: ideally, the light should be monochromatic, the actinometer solution should have the same optical density at this wavelength as the solution under investigation, the solvent should be the same in the two solutions, and the quantum yield for the actinometer should be similar in magnitude to that for the unknown solution. Such an ideal match is rarely achieved! However, values that are reasonably reliable in the second significant figure (e.g. 0.37 ± 0.03) can be obtained in well-designed experiments.

The quantum yield is not directly proportional to the reaction rate constant; rather it is a ratio of rates (intensity of light absorption is a rate), some of which are sums of rates for several processes. Comparison of quantum yields is rarely a good guide to the relative reactivity of the excited states involved. A better guide to reactivity is the rate constant obtained from dividing the reaction quantum yield by the excited-state lifetime (1.15). In a concerted reaction where product is formed directly from the excited state, or in a reaction that goes through intermediates which do not revert to substrate but go on efficiently to product, this rate constant is the primary reaction rate constant that we are seeking as a measure of excited-state reactivity.

$$k = \frac{\phi}{\tau} \qquad (1.15)$$

Alternatively, since τ is the reciprocal of the sum of rate constants for all the processes undergone by the excited state, the reaction rate constant may be estimated if the others (e.g. for phosphorescence and intersystem crossing in the case of a triplet) are known from other studies. The most convenient way, however, of measuring τ for a reaction that can be quenched is to carry out a quantitative quenching study at different quencher concentrations. In the most straightforward systems the results can be fitted to a straight-line plot expressed as (1.16), where ϕ_0 is the quantum yield in the absence of quencher, ϕ is the measured quantum yield at quencher concentration $[Q]$, and k_Q is the rate constant for quenching.

$$\frac{\phi_0}{\phi} = 1 + k_Q\tau[Q] \qquad (1.16)$$

A representative plot is shown in Figure 1.15; this is known as a Stern–Volmer plot, and (1.16) as a Stern–Volmer equation. This method for obtaining reaction rate constants is again a comparative one, since there is competition between the primary reaction step and the quenching process. A value for the quenching rate constant needs to be known, but in many cases this is independent of the substrate and quencher because triplet quenching is controlled by diffusional collision of the two species. So for a particular solvent at a given temperature k_Q values are available in the literature; as an example, for benzene at 25°C the value is 5 g 10^9 l mol^{-1} s^{-1}. Note that this is given to only one significant figure—rate constants obtained by this method should not be treated as being of high accuracy, and I would be suspicious of interpretations that relied on differences in the second significant figure in such values.

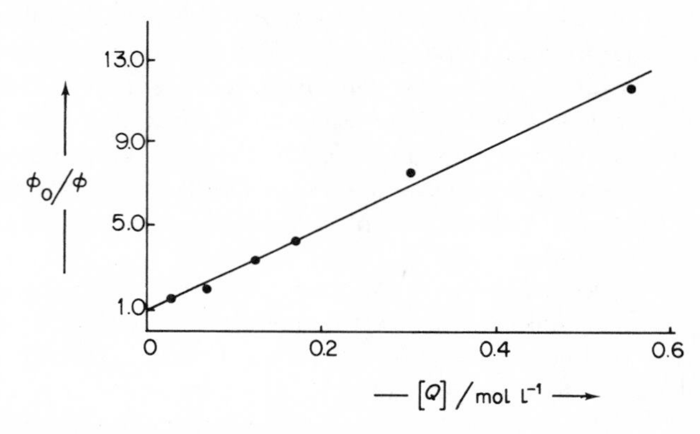

Figure 1.15 A Stern–Volmer quenching plot.

As with other quantitative photochemical studies, it is important to design Stern–Volmer experiments carefully: the quencher should be chosen to ensure that it interacts only with the excited state that is of interest, the extent of reaction should be small enough to ensure that substrate depletion does not affect the intensity of light absorbed, and the concentration of quencher should not be so small that it is significantly depleted by its sensitized reaction. Stern–Volmer plots may turn out to be non-linear, for example because the quencher interacts with more than one excited state on the reaction pathway, or because two different excited states lead to the same chemical product, and such results are of value in unravelling the mechanism.

So far the methods described for measuring excited state lifetime, and hence reactivity, have been indirect methods that rely on a comparison with some standard (e.g. actinometer quantum yield or quenching rate constant) that has already been measured. A direct method for measuring the lifetime of short-lived species produced photochemically is flash photolysis. This is a very important technique in photochemistry, though only the basic ideas as they apply to mechanistic studies are outlined here. In flash photolysis a high concentration of a short-lived species (electronically excited state or chemical intermediate) in a photochemical reaction is generated using a high intensity pulse of radiation of short duration (a 'flash'). When the exciting pulse has died away, which may take only a few picoseconds (1 ps = 10^{-12} s) with laser sources, the system is analysed, usually by monitoring light emission (from an excited state) or light absorption (by any chemical species). The new feature as far as excited state reactions are concerned is that the *absorption* properties of an excited singlet or triplet state are studied, as the excited state is excited further by absorption of a second photon. Each excited state gives rise to a characteristic absorption spectrum, just as the ground state does. Such studies are possible because the exciting flash generates a sufficiently high concentration of the excited state and lasts only a short time compared with the excited-state lifetime.

Whether the excited state is monitored by its emission or by its absorption, decay of the monitoring signal provides a direct measure of the excited-state lifetime. In this way lifetimes can be obtained for non-luminescent states, and for triplet states in fluid solution at room temperature (phosphorescence lifetimes are normally for a rigid glass at 77 K), and k_Q values can be measured for particular substrate/quencher/solvent combinations. The technique applies equally well to chemical intermediates such as radicals, biradicals or radical ions, and in favourable circumstances a sequence of intermediates can be monitored sequentially and shown to be interrelated. Flash photolysis techniques have developed to the stage where they can be used to study extremely rapid photophysical processes, almost as fast as the process of light absorption. They offer a powerful tool for obtaining quantitative information about a reaction pathway to a mechanistic organic photochemist who is in the fortunate position of having access to the necessary equipment.

Finally, it should be pointed out that methods used to study short-lived chemical intermediates in fast thermal reactions may be applicable also to photochemical studies. Radical intermediates, however generated, can be studied by CIDNP (chemically induced dynamic nuclear spin polarization), in which the n.m.r. spectrum of the reaction mixture is recorded during the reaction period. If a substrate is continuously irradiated with ultraviolet/visible light in the cavity of an n.m.r. spectrometer, the resulting n.m.r. spectrum of the substrate/product mixture exhibits intensity variations as compared with the normal spectrum—intensity enhancement, reduction or even reversal (i.e. emission). Note that the spectrum involved is not that of the radical intermediate, but of the stable substrate and product molecules. The difference from a normal spectrum lies in the relative intensities of the individual lines of the spectrum, and the effect arises because radical–radical interactions lead initially to a non-equilibrium distribution of nuclear spin configurations in the product or in regenerated substrate. An analysis of the intensities usually provides information about the spin characteristics of the radical intermediates, and about whether their reaction occurs within or outside the solvent cage in which they were produced. Other very labile intermediates may be generated photochemically in a low-temperature matrix, under conditions in which their lifetime is long enough for conventional spectroscopic measurements to be made. Photochemistry is well suited to such matrix isolation studies, because light absorption is not affected by temperature; the use of a frozen glass at 77 K is not uncommon, and solid nitrogen or argon matrixes at temperatures below 20 K can be used to generate such species as benzyne or cyclobutadiene.

Methods of preparative photochemistry

This is not the place for a lengthy description of appartus that is available for photochemical studies. Rather the main purpose of this final section in the chapter is to point out how very simple it can be to carry out a preparative photochemical reaction in organic chemistry. If visible light, or wavelengths just into the ultraviolet range, is adequate for the reaction, all that is needed is daylight, or an ordinary tungsten-filament light source such as a desk lamp, and a glass container for the sample solution. This works well for such processes as photochemical brominations using molecular bromine, or for photoreactions of coloured organic compounds such as azo-compounds, or for reactions sensitized by coloured dyes.

More commonly a source of ultraviolet radiation is required, and the most widely used sources for preparative work are mercury discharge lamps. A medium-pressure mercury arc (operating pressure up to 10 atmospheres) emits at a variety of discrete ultraviolet wavelengths, notably 313, 366 and 405 nm, as well as in the visible region. A high-pressure mercury arc (operating pressure up to 200 atmospheres) emits more of a continuum of wavelengths across the ultraviolet and visible range. Both types of lamp require a transformer to enable them to operate off mains electricity, and both produce a considerable amount of heat. Cooling is generally provided by circulating water through a double-walled jacket around the lamp, made either of borosilicate (Pyrex) glass for transmission of wavelengths down to about 300 nm, or of quartz for transmission down to about 200 nm. A low-pressure mercury arc gives out largely ultraviolet radiation of wavelength 254 nm, and it produces less heat so that fan cooling is sufficient; these factors may outweigh the disadvantage of the lower power (intensity) of light available. Many other types of ultraviolet and visible lamp may be used in photochemistry, including xenon arcs, sodium lamps, fluorescent-coated mercury arcs, or ultraviolet lasers, but their application to preparative work is generally limited because the output is relatively low-powered.

For many synthetic photoreactions the whole output of a mercury arc can be used. If more selective irradiation is required, for example if the substrate undergoes wavelength-dependent photochemistry or if a product absorbs certain wavelengths and is degraded, the output from the lamp needs to be modified before it reaches the sample. The most convenient way of doing this is to use a filter, either a glass filter that cuts out radiation below a certain wavelength, or a filter solution that acts in a similar way. For example, an aqueous solution of potassium chromate selectively transmits radiation at 313 nm from the ultraviolet output of a medium-pressure mercury arc. A filter solution can be circulated instead of water in the lamp's cooling jacket.

For ultraviolet irradiations care is needed in choosing a solvent—water, alkanes and acetonitrile are transparent and often unreactive towards electronically excited substrates; alcohols and ethers are transparent but are more likely to be reactive; acetone and benzene are sometimes used even though they are not transparent to all relevant wavelengths. It may be necessary to find out by trial and error whether or not an otherwise suitable solvent interferes with the photochemical process. A similar approach can be taken to decide about the need to purge continuously with an inert gas to remove dissolved oxygen.

The basic equipment needed for carrying out photochemical reactions in order to prepare organic compounds is therefore quite simple, and on a laboratory scale quantities up to tens of grams may be handled with ease. Much of the more sophisticated apparatus used in photochemistry is for quantitative, mechanistic or photophysical studies, and the synthetic organic chemist rarely, if ever, needs it!

Further reading

J. A. Barltrop and J. D. Coyle, *Principles of Photochemistry*, Wiley (1978). This book provides a more extensive account of the theoretical and physical aspects of organic photochemistry.

S. L. Mattes and S. Farid, *Science*, vol. 226 (1984), pp. 917–921. A short review of the role of electron transfer in organic photochemical reactions shows how the importance of such processes is becoming more adequately recognized.

J. D. Coyle, R. R. Hill and D. R. Roberts (eds), *Light, Chemical Change and Life*, Open University Press (1982). This collection of short articles provides a complement to any account of the basic principles of photochemistry, setting relevant material in a natural, social, technological or laboratory context.

R. Roberts, R. P. Ouellette, M. M. Muradaz, R. F. Cozens and P. N. Cheremisinoff, *Applications of Photochemistry*, Technomic Publishing Co. (1984). Whilst weak on basic principles, this book gives a good brief overview of biomedical and technological applications of photochemistry, including uses of lasers.

W. M. Horspool, *Synthetic Organic Photochemistry*, Plenum (1984), chap. 9. A short account of apparatus for organic photochemistry is found in this chapter, with leading references to sources of greater detail.

CHAPTER 2

Photochemistry of alkenes and related compounds

Alkenes appear to offer one of the simplest groups of organic compounds as far as photochemistry is concerned, because their electronic structure can be described in terms of only σ and π bonding orbitals, with the corresponding π^* and σ^* antibonding orbitals. However, this appearance is deceptive for monoalkenes, because they turn out to have three low-lying singlet excited states of similar energy. The most intense absorption band has its maximum around 180 nm ($\epsilon \sim 10^4$ l mol^{-1} cm^{-1}) and corresponds to a $\pi \rightarrow \pi^*$ transition; weaker absorption at longer wavelengths has been attributed to a $\pi \rightarrow \sigma^*$ transition; finally, gas-phase spectra show clear evidence of Rydberg absorption leading to a Rydberg state designated as (π,3s). It is tempting to suggest that each of these excited states—(π,π^*), (π,σ^*), and (π,3s) is responsible for different types of photochemical behaviour, but this is difficult to substantiate because those monoalkenes that show the clearest evidence for multiple low-lying states are not easy to study experimentally since their main absorption is beyond the readily accessible ultraviolet range (i.e. is below 200 nm). Alkenes that absorb at longer wavelengths, such as conjugated dienes and polyenes or aryl-substituted alkenes, do not have this drawback, but for these the state diagram is more straightforward and the lowest excited singlet state is clearly (π,π^*). These alkenes offer a fascinating array of photochemistry, and many of the examples in this chapter involve such conjugated systems.

The reactions of alkenes and related compounds are grouped here into nine sections. The first five deal essentially with photoisomerization processes—geometrical isomerization about a carbon–carbon double bond, concerted (electrocyclic) cyclization, concerted shifts (usually of hydrogen) along the π-system, the di-π-methane reaction, and a group of other isomerizations of monoalkenes. Three sections describe photoaddition reactions—acyclic addition, cycloaddition, and the special case of addition to oxygen—and the final section covers alkynes. A feature of alkene photochemistry is that singlet and triplet state reactions are more distinct than for any other major group of compounds. There is a very large difference in energy between a singlet (π,π^*) state and the corresponding triplet state, and intersystem crossing is very inefficient; the triplet photochemistry of alkenes is studied largely with the aid of triplet sensitizers. The large singlet–triplet energy difference and the different spin characteristics lead to different reactions for the singlet state (direct irradiation) and the triplet state (sensitized irradiation) of many alkenes, and even where the reactions are not different in nature, other features diverge such as the relative rates for competing pathways, or the stereochemical course of reaction.

When an alkene absorbs a photon, the (π,π^*) excited state is formed initially in a conformation that retains the geometry of the

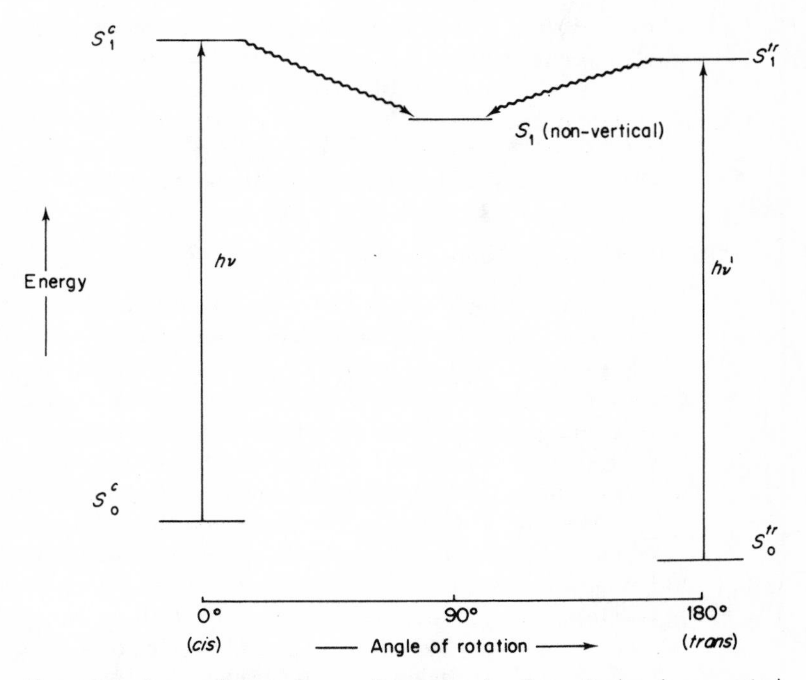

Figure 2.1 Energy diagram for an alkene, showing the vertical and non-vertical singlet excited states.

ground state from which it is formed (the Franck–Condon principle, see p. 15). These conformations are sometimes called the vertical excited states because of their relationship to the ground states on a potential energy diagram or on a state energy diagram (Figure 2.1). However, there is no net π-bonding in the (π,π^*) state and little barrier to rotation about the erstwhile double bond, with the result that relaxation can take place rapidly to give a non-vertical (π,π^*) state; this has lower energy and a different geometry, with an angle of twist of around 90° compared with the planer geometry of the vertical state. The large change in geometry between ground state and relaxed singlet excited state, together with rapid radiationless decay from the relaxed state, accounts for the observation that alkene fluorescence is usually extremely weak, and a similar situation holds for triplet states, so that phosphorescence is also weak. Strong luminescence from alkenes requires some molecular feature that inhibits twisting about the double bond, such as the increased rigidity found in some polycyclic systems.

The loss of stereochemical 'memory' in the non-vertical (π,π^*) excited state implies that stereospecific, concerted reactions of an alkene singlet state may take place from the vertical state. Of particular importance is that the change from *cis* or *trans* geometry to something in between opens up a route for converting one geometrical isomer of an alkene to another, and this is a photoisomerization reaction that will be described in the next section.

Geometrical isomerization

Thermal or catalytic methods for interconverting geometrical isomers of alkenes lead to a thermodynamic equilibrium mixture of isomers. For example, the *cis* isomer of stilbene (PhCH=CHPh) reacts to produce mainly the more stable *trans* isomer when heated strongly or when treated with a proton acid that acts catalytically by the reversible formation of a carbonium ion. A photochemical catalytic method that operates in a related way involves the use of a halogen (usually iodine, so that electrophilic addition to the alkene is not a competing reaction). The halogen molecule is cleaved photochemically to give halogen atoms, and these can add reversibly to the alkene forming an intermediate radical in which bond rotation is possible. Such a process catalysed by bromine is useful for obtaining the more stable *cis*-1, 2-dichloroethylene from the more readily obtained *trans* isomer (2.1).

hν, Br_2 (2.1)

70%

Genuine photochemical isomerization involves an excited state of the alkene, and its great utility is in providing a route to the thermodynamically less stable isomer. On direct irradiation the initially formed vertical excited state (*cis* or *trans*) of an alkene relaxes to the non-vertical state which does not retain stereochemical information in its orthogonal geometry. The non-vertical state can relax to either *cis* or *trans* ground state by radiationless decay, and for many systems there is an approximately equal probability of forming the *cis* or *trans* isomer. This means that the main factor governing the final ratio of isomers is the relative probability of exciting the ground state isomers in the first place, which is measured by the relative absorption coefficients at the wavelength used for irradiation. The commonest situation is that when either isomer of an alkene is irradiated at a particular wavelength the ratio of *cis* or *trans* compound approaches a value that is characteristic for that wavelength. When no further change occurs a 'photostationary state' has been achieved in which the rate of formation of each isomer (from the non-vertical state) equals the rate of its removal (by absorption of light). In practice other reactions of lower efficiency gradually deplete the photostationary state, and many photocyclizations that require the *cis* isomer of an alkene as substrate can be conducted starting with the *trans* isomer.

If the object is to prepare a particular geometrical isomer, the choice of wavelength is decisive, and the use of a wavelength where the more stable isomer absorbs more strongly than the less stable one makes for the effective formation of the less stable isomer. For the stilbenes, the choice of 313 nm ($\epsilon_{cis} = 2300$, $\epsilon_{trans} = 16{,}300$ l mol^{-1} cm^{-1}) gives a photostationary state mixture containing 93% *cis*-stilbene (2.2); *cis,cis*-cyclo-octa-1,3-diene at 248 nm is converted into a mixture richer in the much less stable *cis,trans* isomer (2.3).

hν (313 nm) (2.2)

7% 93%

$h\nu$ (248 nm)

63% 37%

(2.3)

The *trans* isomers of eight-membered cycloalkenes are usually kinetically stable under normal conditions and can be isolated. Those of seven- or six-membered cycloalkenes cannot be isolated (except as complexes with copper(I) species), but they are thought to be intermediates in a number of photochemical addition or cycloaddition reactions, as described in later sections of this chapter. Although cyclo-octadiene shows the behaviour expected on the basis of monoalkenes, there are two complicating factors in considering the geometrical isomerization of conjugated dienes and polyenes. First, ground-state conformations of the diene about the central single bond are important, because this bond has substantial double bond character in the excited state, and the excited states obtained from rapidly interconverting ground-state conformers may themselves be non-interconvertible on the time-scale of subsequent reactions (2.4). Secondly, the relaxed, non-vertical excited states for conjugated systems may not have the orthogonal geometry characteristic of such states for the monoalkenes, but may instead have an angle of twist smaller than 90°. The outcome is that, whereas conjugated systems do exhibit a wavelength-dependent composition for their photostationary states, the theoretical basis for understanding the variation is quite complex.

$h\nu$

(2.4)

Photochemical *cis–trans* isomerization in a conjugated polyene system is thought to be the crucial primary process in vision. The visual pigment (rhodopsin) is derived from 11-*cis*-retinal by reaction of the aldehyde group with an amino substituent in a protein (opsin). There is considerable distortion in the geometry of this chromophoric group anyway, because of the spatial requirements of the protein structure, but it does seem that the key event in recognition of a photon by the visual system is photoisomerization about the 11,12-double bond to give a derivative of all-*trans*-retinal (2.5).

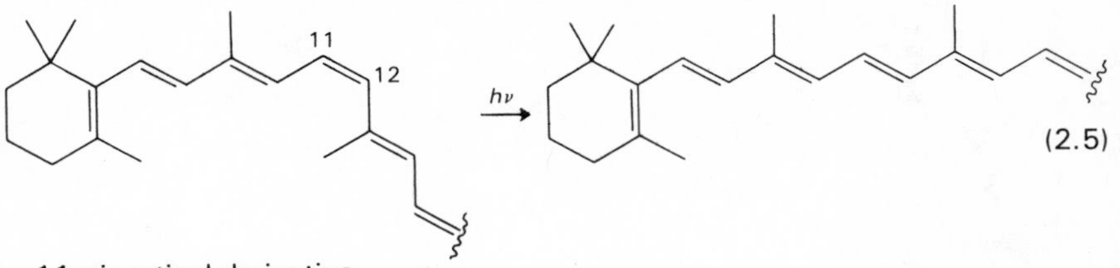

11 12 $h\nu$ (2.5)

11-*cis* retinal derivative

Photochemical isomerization of the geometrical isomers of alkenes can also take place through triplet (π,π^*) states, since the lowest energy triplet conformation, like that for excited singlet (π,π^*) states, has a twisted geometry intermediate between that of the *cis* and *trans* isomers. The non-vertical triplet state relaxes to ground state *cis* or *trans* alkene with roughly equal probability, but in a sensitized reaction the absorption coefficients of the alkenes at the wavelength used are not relevant in determining the photostationary state composition. A triplet sensitizer acts by absorbing light, undergoing vibrational relaxation and intersystem crossing to its triplet state, and then taking part in an energy transfer process with the alkene, generally by an exchange mechanism (see Chapter 1). The key property is the value of the triplet energy of the sensitizer in relation to that of the energy acceptor. To be effective the sensitizer triplet energy should be greater than that of the alkene, and since the alkene isomers have different triplet energies, there will be a (small) range of sensitizers for a particular alkene whose triplet energy lies between that of the *cis*-alkene and that of the *trans*-alkene. In such a situation the alkene isomer with the lower triplet energy is sensitized more efficiently then the isomer with the higher triplet energy, and the result is a photostationary state rich in the latter isomer. Even when the choice of sensitizer cannot be made on the basis of experimentally determined alkene triplet energies, the use of a triplet sensitizer to obtain the less stable isomer may still be effective. This is seen in the preparation of the very hindered *cis* isomer of 3,3-dimethyl-1-phenylbut-1-ene (2.6) using an aromatic ketone as sensitizer, and in the formation of a *trans*-cyclo-octene (2.7) with toluene as sensitiser.

There are applications of photosensitized *cis–trans* isomerization other than the preparation of one isomer from another. By using

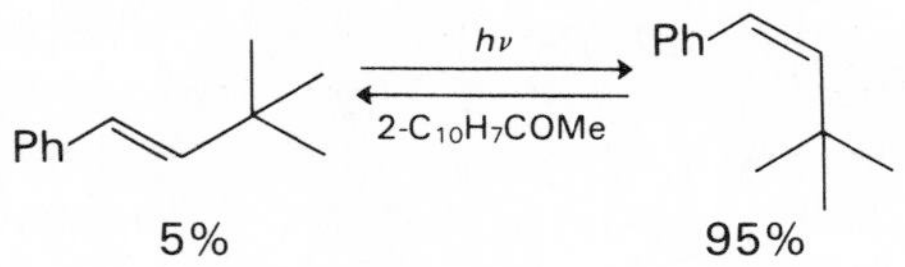

Ph 5% $\underset{2\text{-}C_{10}H_7COMe}{\overset{h\nu}{\rightleftarrows}}$ Ph 95% (2.6)

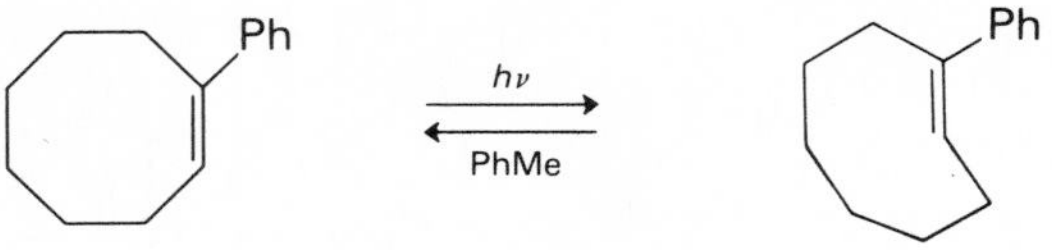

Ph $\underset{PhMe}{\overset{h\nu}{\rightleftarrows}}$ Ph (2.7)

well-studied systems such as stilbene or penta-1,3-diene, the variation of the photostationary state ratio of isomers with the triplet energy of the sensitizer can be plotted, and this allows an estimate to be made of the triplet energy for any new compound that can be used as a triplet sensitizer. A related use of photo-sensitized alkene isomerization involves measuring the initial relative rate of sensitized conversion of *cis*-penta-1,3-diene to its *trans* isomer in order to determine the efficiency of intersystem crossing for a particular sensitizer, relative to that for a compound whose intersystem crossing efficiency is known.

It has been assumed so far that the sensitizer acts by an energy-transfer mechanism, but in some cases other modes of interaction may occur. It is possible that electron transfer takes place to give the radical anion or the radical cation of the alkene, which is the species that subsequently isomerizes. This is likely to be the case in the chlorophyll-sensitized isomerization of vitamin A acetate, which is used commercially to obtain the required all-*trans* isomer (2.8) from the mixture of isomers resulting from the synthesis. Unlike triplet-sensitized reactions, electron-transfer isomerizations frequently lead to a predominance of the most thermodynamically stable isomer.

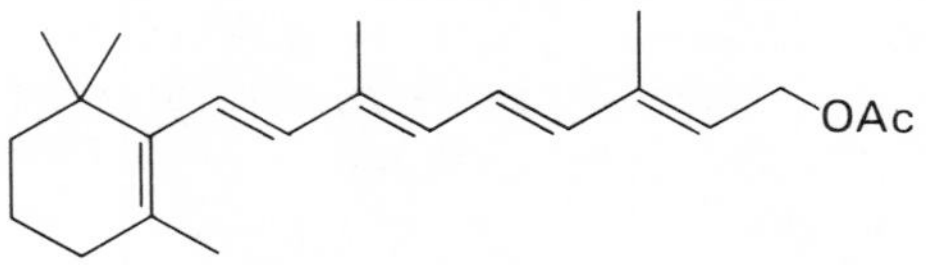

OAc (2.8)

Electrocyclic processes

Electrocyclic processes are one major group of concerted reactions for alkenes, that is, reactions in which all the bond-making and bond-breaking events occur in a single reaction step. In electrocyclic ring-closure a new σ-bond is formed between the terminal atoms of a conjugated π-system; electrocyclic ring-opening is the reverse of this. Simple examples are the interconversion of buta-1,3-diene and cyclobutene (2.9), hexa-1,3,5-triene and cyclohexa-1,3-diene (2.10), and octa-1,3,5,7-tetraene and cyclo-octa-1,3,5-triene (2.11), involving 4-, 6- and 8-electron systems, respectively (the number of electrons undergoing major reorganization).

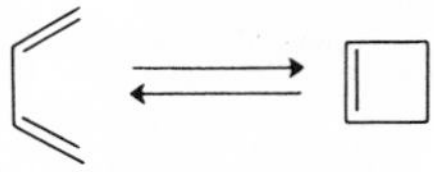

(2.9)

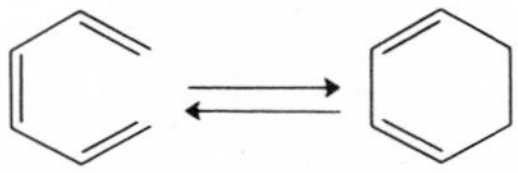

(2.10)

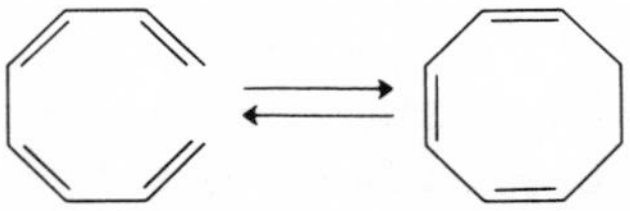

(2.11)

Such reactions may occur thermally or photochemically, and the differences between the two normally show up in two ways. First, in a thermal reaction the direction of change will be towards the equilibrium position, favouring the more thermodynamically stable compound, whereas in a photochemical reaction the direction of change will be towards a photostationary state that favours the compound with the lower absorption coefficient at the wavelength of irradiation. It is therefore normal for conjugated dienes to be converted efficiently into cyclobutenes using wavelengths that are absorbed by the diene but not by the cycloalkene (2.12).

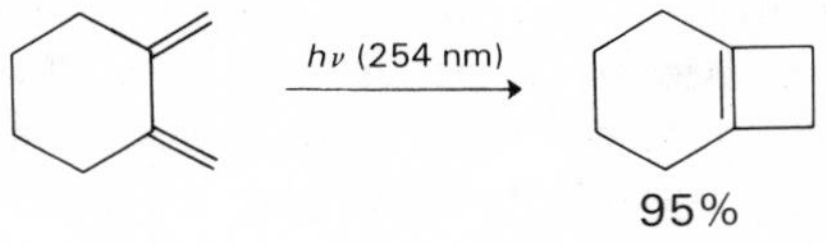

$h\nu$ (254 nm)

95%

(2.12)

The second difference between thermal and photochemical electrocyclic processes is seen in the stereochemical course of reaction; both types of process are stereospecific, but for a given system of electrons the thermal and photochemical specificities are in opposite senses. The relevant stereochemical feature is the relationship between the substituents on the atoms which are the termini of the π-system in the ring-opened structure. The two possibilities for a *trans,trans*-1,4-disubstituted buta-1,3-diene are shown in (2.13). Formally the substituents in the cyclobutene are derived from those in the butadiene (or vice-versa) either by a conrotatory mode of reaction, in which rotations occur in the same sense (clockwise or anticlockwise) about the axes of the double bonds, or by a disrotatory mode, in which rotations occur in opposite senses (one clockwise, the other anticlockwise) about these axes.

disrotatory closure

conrotatory closure

(2.13)

The observed stereochemistry of electrocyclic reactions is codified in the following table, where the total number of electrons (*N*) involved in the major bonding changes is expressed as a multiple (4*n*), or not a multiple (4*n* d 2), of four.

N	Thermal reaction	Photochemical reaction
4*n*	Conrotatory	Disrotatory
4*n* + 2	Disrotatory	Conrotatory

This is a specific form of the more general Woodward–Hoffmann rules, which apply to a wide range of concerted reactions. These rules have been treated theoretically in a variety of ways, and they have proved to be valid in many classes of reactions, provided they are applied in a 'permissive' way, that is, the rules predict the preferred stereochemical course of reaction, assuming that the processes are viable on other grounds such as consideration of thermodynamic factors. *trans–trans*-Octa-2,4,6-triene provides a nice illustration of the opposite sense of specifity for thermal and photochemical reactions: on heating the triene, it cyclizes to give *cis*-5,6-dimethylcyclohexa-1,3-diene (2.14) by a disrotatory mode of reaction in a 6-electron system, but on irradiation the *trans* isomer is formed (2.15) by a conrotatory closure. In the simple 4-electron systems, photochemical reaction is normally a disrotatory ring-closure and thermal reaction a conrotatory ring-opening, and this provides a route for *cis–trans* isomerization of a diene (2.16).

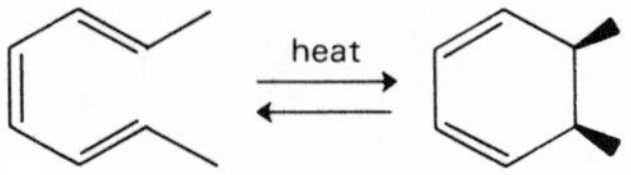

heat (2.14)

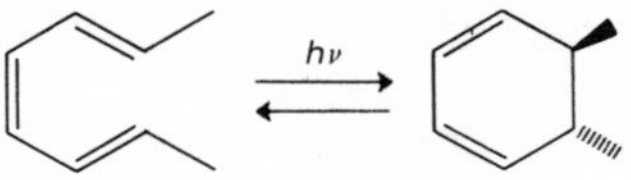

$h\nu$ (2.15)

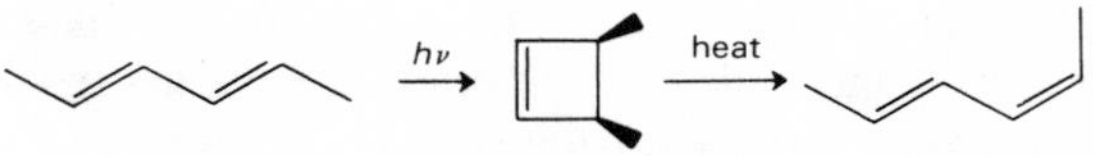

$h\nu$ heat (2.16)

Photochemical electrocyclic ring-closure in a 4-electron system works well for many acyclic dienes (2.17) and related cyclic systems (2.18). The situation with conjugated trienes is more complex, and they can act as 6-electron systems (2.19) leading to cyclohexa-1,3-dienes, or as 4-electron systems (2.20) giving cyclobutenes. In addition they can undergo other photochemical reactions such as geometrical isomerization about the central double bond (which must be *cis* if a 6-electron electrocyclic ring-closure is to take place).

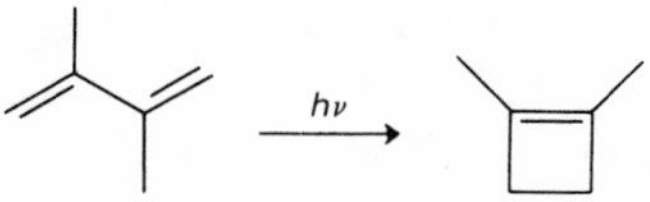

$h\nu$ (2.17)

80%

O $h\nu$ O (2.18)

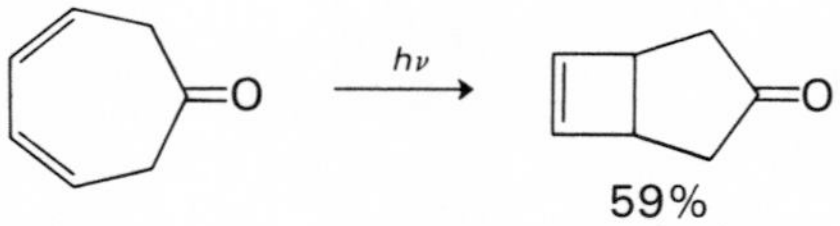

59%

$h\nu$ (2.19)

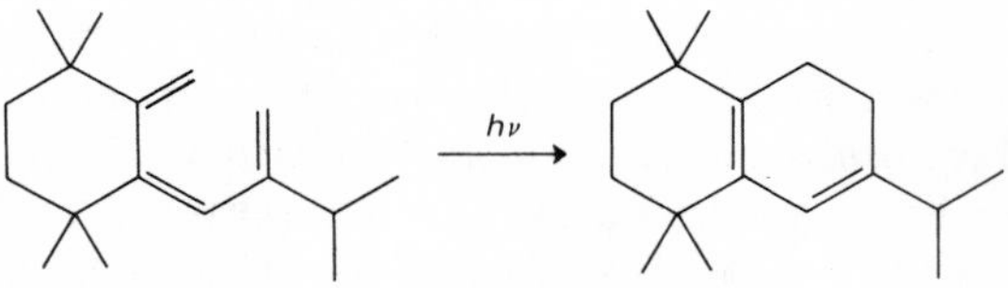

$h\nu$

50% (2.20)

(2.21)

s-trans, s-trans *s-cis, s-trans* *s-cis, s-cis*

An important controlling factor is the conformation of the triene about the two single bonds that link the three double bonds (2.21); an *s-cis–s-cis* conformation is required for 6-electron ring-closure, but 4-electron closure can occur also in an *s-cis–s-trans* conformation. The different conformations interconvert rapidly in acyclic trienes, but because each has different absorption characteristics, the wavelength of irradiation can influence the course of reaction dramatically. 2,5-Dimethylhexa-1,3,5-triene undergoes efficient *cis–trans* isomerization at 254 nm (where the major absorbing species is the *s-trans–s-trans* conformation), but it cyclizes to a cyclohexadiene (2.22) using 313 nm radiation.

$h\nu$ (254 nm)

80%

(2.22)

$h\nu$ (313 nm)

88%

An important group of conjugated diene/triene systems are those in the vitamin D series. The key reactions in the commercial manufacture of vitamin D (and probably also in its formation in skin exposed to daylight) are a photochemical, conrotatory electrocyclic ring-opening in the provitamin, and a thermal 1,7-shift of hydrogen in the previtamin so formed (2.23). High conversions to the vitamin are not normally possible because all three species absorb appreciably at the

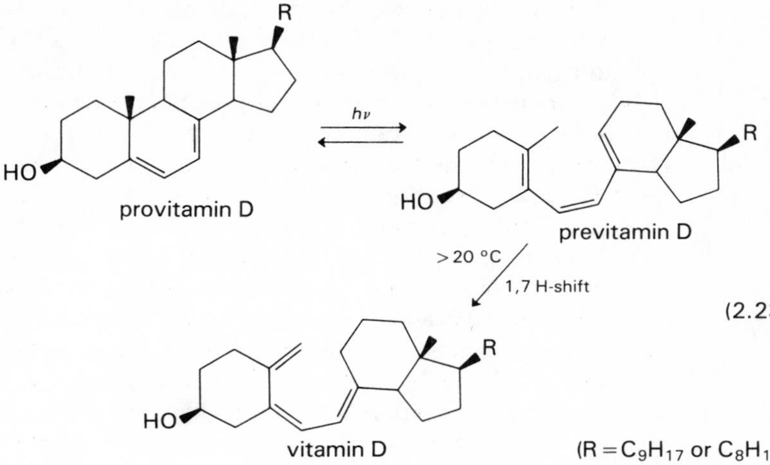

R

HO provitamin D $\xrightarrow{h\nu}$ HO previtamin D R

>20 °C 1,7 H-shift

(2.23)

R

HO vitamin D

($R = C_9H_{17}$ or C_8H_{17})

wavelengths used, and the previtamin, or the vitamin itself if reaction is conducted at a temperature that allows its formation, gives rise to additional products. The use of a selective, monochromatic source of radiation, rather than the normal ultraviolet lamp producing a range of wavelengths, may help to increase the conversion and yield.

Sigmatropic shifts

Sigmatropic shifts, like electrocyclic reactions, are examples of the general class of concerted reactions known as 'pericyclic reactions', in which the reorganization of the bonding electrons occurs in a continuous cyclic array. In a sigmatropic shift, a sigma bond migrates with respect to a system of π-electrons, and as a result of the shift there is a switching of double and single bonds in the π-system. An example is the conversion of hexa-1,3-diene into hexa-2,4-diene (2.24), in which a hydrogen atom moves from carbon-5 to carbon-1.

1 2 3 4 5 6 H ⇄ H (2.24)

This example is said to be a shift of order (1,5), or a 1,5-shift, which means that one end of the sigma bond is involved with just one atom (i.e. it remains attached to this atom, the hydrogen) and the other end of the sigma bond migrates over five atoms inclusive.

In principle, sigmatropic shifts are reversible and may occur either photochemically or thermally; the direction of change is governed by thermodynamic factors (thermal reaction) or light-absorption characteristics (photochemical reaction). Whether or not a particular reaction is seen to occur thermally or photochemically is related to electronic factors and the stereochemical course of the reaction. The most important stereochemical feature is whether the migrating atom or group remains bonded to the same face of the π-system (a suprafacial shift), or whether it becomes bonded to the face opposite its starting position (an antarafacial shift); in many systems, particularly those involving five or fewer atoms in the π-system, geometrical constraints act against an antarafacial shift, and so suprafacial shifts are more common. A second stereochemical feature is the configuration at the migrating group (if it is other than a single atom)—is this configuration retained, or is there inversion? In the examples used in this account there is no configurational question of this kind, and for these reactions to occur photochemically in a suprafacial manner there must be a multiple of 4 electrons involved. Conversely, thermal suprafacial shifts of single atoms are found to involve numbers of electrons that are not a multiple of 4. For antarafacial shifts, when they are observed, the rules are reversed—$(4n + 2)$ electrons in a photochemical reaction, $4n$ electrons in a thermal reaction.

The outcome of all this for photochemical sigmatropic shifts is that those most commonly observed are of order (1,3) or (1,7); these involve 4 or 8 electrons, respectively, and occur in a suprafacial manner. Examples of photochemical 1,3-shifts of hydrogen are found for monoalkenes (2.25) and for conjugated dienes (2.26). In the case of dienes a 1,3-shift is favoured over a 1,5-shift, because for the latter to occur photochemically it would have to take place in an antarafacial manner. Note that in both examples the direction of

$h\nu$ → 85% (2.25)

$h\nu$ → 60% (2.26)

CF_3 CF_3 / CF_3 CF_3 $\xrightarrow{h\nu}$ CF_2 CF_3 F / CF_3 CF_3 (2.27)

100%

CN $\xrightarrow{h\nu}$ CN (2.28)

55%

photochemical change favours the less highly absorbing compound.

Atoms other than hydrogen may migrate in photochemical sigmatropic shifts, such as fluorine (2.27) or carbon (2.28). When a carbon group is involved it is often not possible to tell whether its configuration is retained or inverted (in this example the migrating group is a non-chiral -CH_2R unit), although retention of configuration has been proved in systems that do have a chiral migrating group.

Photochemical 1,7-shifts of hydrogen are observed for conjugated trienes, especially when two (2.29) or three (2.30) of the double

$\xrightarrow{h\nu}$ (2.29)

100%

$\xrightarrow{h\nu}$ (2.30)

CN CN

~ 100%

bonds are constrained in a ring. The function of the ring is to hold the two ends of the π-system in reasonably close proximity; in unconstrained, acyclic trienes other reactions often occur more readily than a 1,7-shift. The second, thermal stage of the vitamin D reaction (2.23) involves a 1,7-shift of hydrogen, but the geometry of previtamin D allows it to adopt quite readily a conformation in which an antarafacial shift can take place. Antarafacial shifts are observed in five-atom systems when there are structural features that favour an *s-cis* conformation of the diene unit and a twisting from planarity; this means

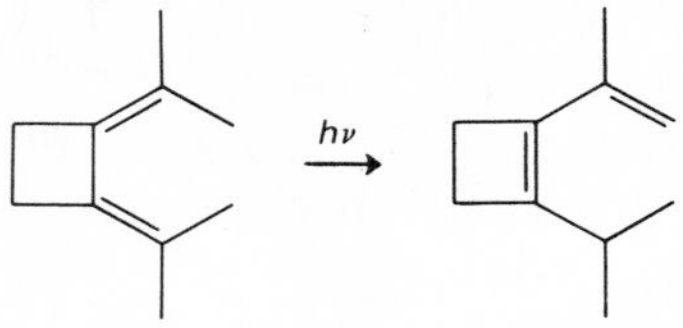

(2.31)

86%

that 1,5-shifts may be observed photochemically in a few instances (2.31).

Di-π-methane reaction

Irradiation of 1,4-dienes or 3-phenylalkenes often leads to vinylcyclopropanes or phenylcyclopropanes, respectively, by processes (2.32) that can be thought of as 1,2-shifts of a vinyl or phenyl group accompanied by ring-closure. A formal mechanism can be drawn (2.33) that starts with bonding between atoms 2 and 4 of the 1,4-diene unit, then cleavage of the 2,3-bond and formation of a 3,5-bond. Even if this is not the pathway followed, it provides a useful

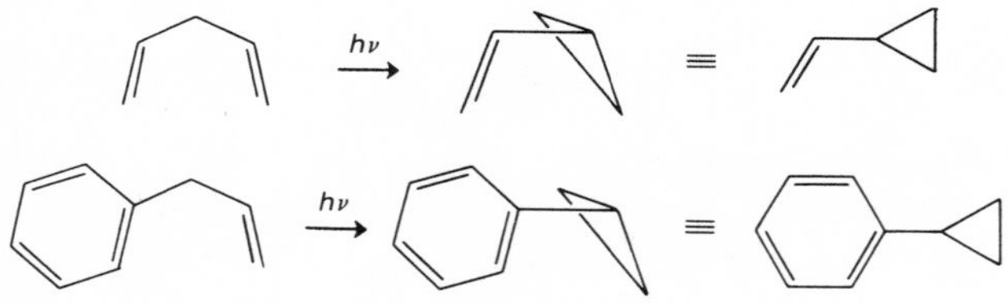

(2.32)

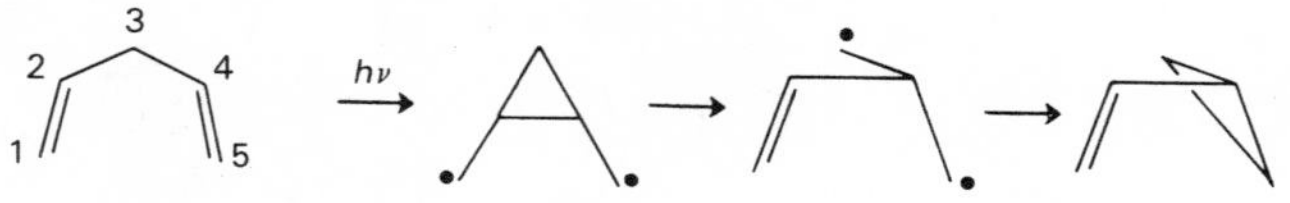

(2.33)

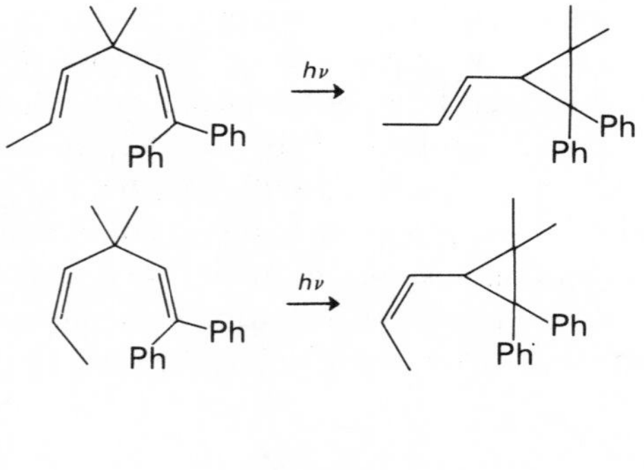

(2.34)

aid to visualizing the products of a di-π-methane reaction, which is not always easy in cyclic systems. Biradical intermediates have not been detected directly, but it does seem that the rearrangement is not a concerted reaction, despite the observed stereospecificity at carbon-1 (2.34), carbon-3 and carbon-5.

For acyclic 1,4-dienes (2.35) or 3-phenylalkenes (2.36) the reaction is normally brought about through the singlet excited state

Ph Ph CN $\xrightarrow{h\nu}$ Ph Ph CN (2.35)

82%

NC CN $\xrightarrow{h\nu}$ CN CN (2.36)

64%

obtained by direct irradiation; the triplet state of such a system is very readily deactivated by rotation about a double bond, and so the triplet di-π-methane process is very inefficient. This situation is reversed when each of the double bonds is constrained in a six-membered, or smaller, ring. With such compounds the di-π-methane reaction is observed mainly from the triplet state on sensitization (2.37, 2.38). The reasons for this difference are that twisting

$\xrightarrow[Me_2CO]{h\nu}$ (2.37)

61%

Ph Ph $\xrightarrow[\text{sensitizer}]{h\nu}$ Ph Ph (2.38)

about the double bond is much less favourable for cyclohexenes or smaller-ring cycloalkenes, and the singlet state often has other efficient reactions that it can undergo, such as electrocyclic ring-opening.

Other photoisomerizations

In this section are found a variety of photorearrangement reactions of alkenes, dienes and polyenes that do not fit into the four mechanistic categories described so far. Accompanying the electrocyclic ring-closure of conjugated dienes is an alternative photocyclization giving a bicyclo[1.1.0]butane (2.39); when the diene is constrained in an *s-trans* conformation (2.40) the bicyclobutane may be the major product because electrocyclic reaction requires an *s-cis* conformation. There is an analogous reaction for conjugated trienes, leading to a bicyclo[3.1.0]hexene (2.41), and this requires an *s-cis, s-trans*

$h\nu$ (2.39)

R R $h\nu$ (2.40)

78%

Ph Ph Ph $h\nu$ Ph Ph Ph (2.41)

76%

conformation in the triene substrate (see equation 2.21), highlighting once again the importance of ground-state conformations in directing the course of photochemical reaction for conjugated polyene systems.

A number of photochemical 1,3-shifts are known that cannot occur by a concerted (sigmatropic) mechanism, such as the sensitized rearrangement a 3-vinylcyclopropene to a cyclopentadiene (2.42). There are also a group of reactions for which the term 'bicycle'

Ph Ph Ph $\xrightarrow{h\nu}$ Ph Ph Ph

87% (2.42)

$\xrightarrow{h\nu}$

87% (2.43)

$\xrightarrow{h\nu}$

75% (2.44)

rearrangements has been coined (2.43); to account for these a number of sequential shifts can be envisaged through biradical intermediates. Many of these reactions require quite exotic starting materials, and they are studied because of their mechanistic fascination.

The di-π-methane reaction results in a 1,2-shift in a 1,4-diene unit, but such shifts sometimes occur in monoalkenes (2.44), and the mechanism must be different. The substrates are usually tetrasubstituted ethylenes, and it is suggested that the reactive excited state is a Rydberg singlet state, which rearranges initially to give a carbene. Support for such a mechanism comes from the structures of products obtained from 1,2-dimethylcyclohexene (2.45), which are

$\xrightarrow{h\nu}$

24%

(2.45)

21% 38%

consistent with a carbene precursor. Shifts of hydrogen across more than three carbon atoms are sometimes observed in the sensitized reactions of alkenes, especially cyclic alkenes in which the relevant atoms are held closer together. In this way cyclo-octene provides a series of bicycloalkanes on sensitization with mercury (2.46).

$$\text{cyclo-octene} \xrightarrow[\text{Hg}]{h\nu} \text{bicyclo[5.1.0]octane} + \text{bicyclo[4.2.0]octane} + \text{bicyclo[3.3.0]octane} \qquad (2.46)$$

Addition reactions

One of the most characteristic types of ground-state reaction for alkenes is electrophilic addition, often involving a proton acid as addend or catalyst. In the excited state similar reactions can occur, with water, alcohols or carboxylic acids as commonly encountered addends. However, there is a variety of photochemical mechanisms according to the conditions or substrate used. In a few instances it is proposed that the electronically excited state is attacked directly by a proton from aqueous acid, for example when styrenes are converted to 1-arylethanols (2.47); the rate constant for such attack is estimated to be eleven to fourteen orders of magnitude greater than that for attack on the ground state, and the orientation of addition is that expected on the basis of relative carbonium ion stabilities (Markownikov addition).

Addition to alkenes can be sensitized by both electron-donors and electron-acceptors, and it is most likely that the reactive species is the alkene radical anion or the alkene radical cation, respectively. 1,1-Diphenylethylene can be converted to the Markownikov addition product with methanol (2.48) using the electron-donating sensitizer 1-methoxynaphthalene; no added proton acid is needed. Using

$$\text{ArCH=CH}_2 \xrightarrow[H_2O,\ H_2SO_4]{h\nu} \text{ArCH(OH)CH}_3 \qquad (2.47)$$

$$Ph_2C{=}CH_2 + MeOH \xrightarrow[1\text{-}C_{10}H_7OMe]{h\nu} Ph_2C(OMe)CH_3 \qquad (2.48)$$

50–80%

methyl *p*-cyanobenzoate as electron-acceptor sensitizer, the anti-Markownikov product is formed (2.49) via the alkene radical cation. In this way the orientation of addition can be controlled by the choice of sensitizer, and mild, non-acidic conditions can be employed. The photochemical addition may also succeed with very weakly nucleophilic compounds such as 2,2,2-trifluoroethanol (2.50), which often do not add readily to alkenes under thermal, acid-promoted conditions.

$$Ph_2C{=}CH_2 + MeOH \xrightarrow[p\text{-}NCC_6H_4CO_2Me]{h\nu} Ph_2CHCH_2OMe \quad 50\% \qquad (2.49)$$

$$\text{1-phenylcyclohexene} + CF_3CH_2OH \xrightarrow[1\text{-}C_{10}H_7OMe]{h\nu} \text{1-Ph-1-}OCH_2CF_3\text{-cyclohexane} \quad 73\% \qquad (2.50)$$

The anti-Markownikov orientation of addition in the presence of electron-acceptor sensitizers applies also to intramolecular reaction, and 5,5-diphenylpent-4-en-1-ol gives a tetrahydrofuran (2.51) when irradiated in solution with 9,10-dicyanoanthracene, whereas its thermal reaction under proton-acid catalysis leads to 2,2-diphenyltetrahydropyran by Markownikov addition. Sometimes an added sensitizer is not required, if the alkene itself can act as a good electron-donor or electron-acceptor, and this is likely to be the reason why 1-(*o*-methoxyphenyl)propene adds photochemically to acetic acid (2.52), whereas 1-phenylpropene does not.

$$Ph_2C{=}CHCH_2CH_2CH_2OH \xrightarrow[C_{14}H_8(CN)_2]{h\nu} \text{2-}(Ph_2CH)\text{-tetrahydrofuran} \qquad (2.51)$$

$$\text{1-(o-MeOC}_6H_4)\text{propene} \xrightarrow[AcOH]{h\nu} o\text{-}MeOC_6H_4CH(OAc)CH_2CH_3 \quad \sim 100\% \qquad (2.52)$$

A further mechanism for photoaddition that applies to cyclohexenes or cycloheptenes begins with formation of the highly reactive *trans* isomer of the cycloalkene. In this way 1-methylcycloheptene gives an ether on irradiation in methanol (2.53), and 1-methylcyclohexene an acetate with acetic acid (2.54). In both cases a

+ MeOH $\xrightarrow[\text{xylene}]{h\nu}$ OMe 93% (2.53)

+ AcOH $\xrightarrow[\text{xylene}]{h\nu}$ OAc 93% (2.54)

high-energy triplet sensitizer is employed, both to capture the available radiation more effectively, and to avoid competing side-reactions that occur from the singlet excited state of the alkenes. The addend adds to the *trans*-cycloalkene, and this reaction sometimes requires a trace of added mineral acid to ensure its success. In the presence of mineral acid but without a nucleophilic addend, reaction can lead to an isomer of the alkene with an exocyclic double bond (2.55) that cannot undergo reverse reaction because it does not form a strained *trans*-isomer as the cycloalkene does.

The restriction of this mechanism to cycloalkenes with a six- or seven-membered ring enables selective reaction to be achieved with a non-conjugated diene such as limonene (2.56), for which thermal reaction would occur at both double bonds.

$\xrightarrow[\text{Et}_2\text{O, H}^+]{h\nu\text{, xylene}}$ 95% (2.55)

$\xrightarrow[\text{aq. Bu}^t\text{OH, H}^+]{h\nu\text{, xylene}}$ OH 89% (2.56)

Cyclopentenes behave differently and often act through radical mechanisms; this can lead to photoreduction to cyclopentanes, or photoaddition of the kind exemplified by norbornene and propan-2-ol (2.57). The photoadduct in this process is linked through the carbon atom of the alcohol, and not the oxygen atom. A related addition to acetonitrile (2.58) takes place when norbornene is irradiated in the presence of a silver(I) compound. It is likely that a metal complex of the alkene is the real irradiation substrate, and the same may be true for copper(I)-promoted additions of haloalkanes to electron-deficient alkenes (2.59). When dichloromethane is used in such a reaction the product can be reduced electrochemically to a cyclopropane (2.60), which is of value because the related thermal addition of CH_2I_2 to alkenes in the presence of copper does not succeed with electron-poor compounds.

OH + ; $h\nu$ → OH 33% + ( )$_2$ 14% (2.57)

+ MeCN $\xrightarrow[AgOSO_2CF_3]{h\nu}$ CN 93% (2.58)

CN + Br $\xrightarrow[CuCl]{h\nu}$ Br, CN 62% (2.59)

CN + CH_2Cl_2 $\xrightarrow[CuCl]{h\nu}$ Cl, Cl, CN 90% $\xrightarrow{2e^-}$ CN 83% (2.60)

Cycloaddition reactions

In a cycloaddition reaction, two sigma bonds are formed to give a new ring of atoms, and one classification of these reactions is based on the number of ring atoms that derive from each of the reacting units. One of the simplest classes are (2 + 2) cycloadditions, such as the reaction of two alkenes to form a cyclobutane (2.61). In a (4 + 2) cycloaddition a new six-membered ring is produced from a four-atom unit and a two-atom unit, such as a conjugated diene and an alkene (2.62). Photochemical (2 + 2) cycloadditions are of very

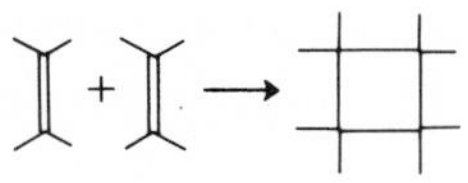

(2.61)

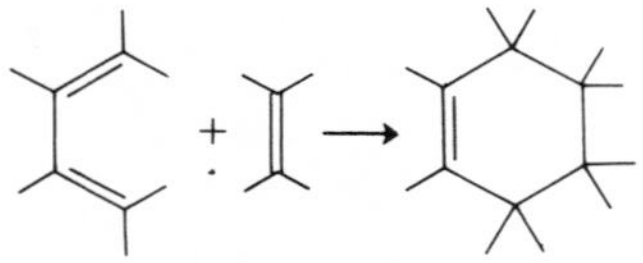

(2.62)

widespread occurrence, and those involving two alkenes are of considerable interest, both for their varied mechanisms and for their usefulness in synthesis. Two of the factors that make these reactions so successful are that thermodynamic barriers to forming cyclobutanes are overcome by the much higher internal energy of an electronically excited state, and that the cyclic products are almost always less highly absorbing (and do not have low-lying triplet states), so that the reverse reaction is not important.

Alkenes sometimes form cyclobutane-dimers by direct irradiation (2.63), and this occurs by way of an excited singlet state. When *cis*- or *trans*-but-2-ene is irradiated at reduced temperature as a neat liquid, the dimers, which accompany *cis–trans* isomerization and a 1,3-shift of hydrogen to give but-1-ene, are formed stereospecifically in a manner that preserves the ground-state geometry (2.64), although

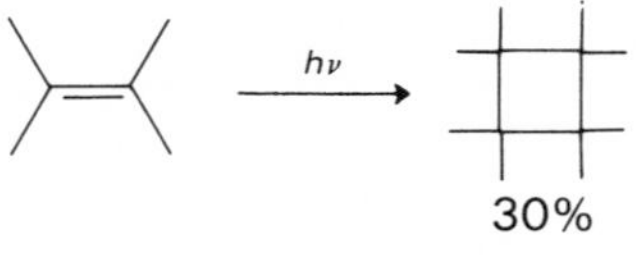

$h\nu$ 30% (2.63)

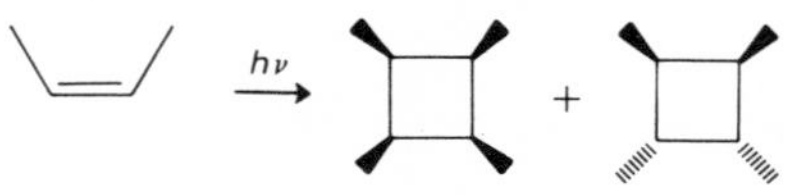

$h\nu$ (2.64)

the stereoselectivity is much reduced for dilute solutions in an alkane solvent. These results imply that the reactive excited state can retain a stereochemical memory of its original ground state, and although a vertical (π,π^*) singlet is a strong candidate, a Rydberg excited state has been proposed as responsible for the photodimerization, attacking a second, ground-state molecule of alkene.

The preservation of alkene geometry in these concerted cycloadditions implies a mode of reaction in which, for each alkene unit, the two new bonds are formed to the same face of the π-bond. This mode is called 'suprafacial–suprafacial' cycloaddition, and is one of

hν → + (2.65)

several possible modes that are dealt with in complete descriptions of thermal and photochemical cycloadditions. For the purposes of this section, in which the examples are mainly non-concerted (2 + 2) cycloadditions, a further consideration of the stereochemical course of concerted reactions is not necessary.

Triplet sensitization is a very effective way of promoting photocycloaddition of fairly rigid alkenes for which *cis–trans* isomerization is inhibited; the sensitizer is required because alkenes generally do not undergo efficient intersystem crossing from the singlet state generated by direct excitation. Cyclopentene gives tricyclic dimers (2.66) by this method, and norbornadiene leads to an internal cycloadduct called quadricyclane (2.67). Substituted versions of the norbornadiene reaction have been proposed as systems for the collection and

hν, acetone → 56% (2.66)

hν, PhCOMe → 57% (2.67)

storage of solar energy, since the tetracyclic products are high in internal energy, and this energy can be released in a controlled manner using a metal catalyst that causes reversion to the original diene. The process illustrates the great value of intramolecular (2 + 2) photocycloadditions in generating polycyclic 'caged' compounds, and another example is seen in the reaction of a dimer of cyclopentadiene (2.68).

$h\nu$ acetone

62%

(2.68)

Conjugated dienes take part readily in triplet-sensitized photodimerization, and the products obtained from buta-1,3-diene (2.69) include a (4 + 2) adduct as well as stereoisomeric (2 + 2) adducts. The reaction is non-concerted, and a rationalization for the products is provided on the basis of the formation of a biradical intermediate as shown (which is the most stable of the three possible biradicals that might be formed in the first step), by the attack of triplet diene on ground-state diene. Cross-addition takes place in some systems, such as myrcene (2.70) where a triplet diene group attacks the alkene within the same molecule; direct irradiation of myrcene gives mainly

$h\nu$ Ph_2CO

(2.69)

$h\nu$ sensitizer

100%

$h\nu$

(2.70)

the cyclobutene that arises by electrocyclic ring-closure of the conjugated diene unit.

By far the most widespread type of photocycloaddition in synthetic applications employs the alkene unit of an α,β-unsaturated carbonyl compound as one of the addends. Simple examples of such reactions are the dimerization of cyclopent-2-enone (2.71), and the photoaddition of 2-methylpropene to cyclohex-2-enone (2.72). The function of the carbonyl group is to bring the absorption of the alkene into

hν → 55% + 45% (2.71)

+ hν → 54% (2.72)

a more readily accessible range of the ultraviolet, and to promote intersystem crossing so that an added triplet sensitizer is not needed. The examples given do not show explicitly that these triplet-state reactions are not stereospecific (although steric effects may lead to reasonably high stereoselectivity in some cases), but they do illustrate the point that the orientation of addition cannot always be readily rationalized or predicted. It seems that the formation of exciplexes along the reaction pathway may play a significant role in determining the preferred orientation.

The photocycloaddition of alkenes to α,β-unsaturated carbonyl compounds is one of the most widely applied photochemical reactions in synthetic organic chemistry. This is because it provides a ready route to compounds with four-membered rings (including some systems with fairly complex polycyclic structures), the products can be designed with a variety of substituents that allow further chemical transformations to be carried out, and the reactions usually proceed in good yield and, in some cases, with a fair degree of stereoselectivity. Reaction (2.73) shows the key stage in the synthesis of the natural product β-bourbonene; the cyclopentene carries a protected ketone group, and the unsaturated lactone a protected alcohol group, which are used in subsequent stages of the synthesis. An intramolecular example (2.74) illustrates the formation of polycyclic structures, in this case opening up a route to the system of three fused cyclopentane rings found in some natural products. Both this example and another reaction (2.75), also used in making a natural

$h\nu$ OCOBu^t → OCOBu^t 25% (2.73)

$h\nu$ → MeOH → OMe H >90% (2.74)

CO_2Me CO_2Et $h\nu$ → CO_2Me CO_2Et 67% (2.75)

OH + $h\nu$ → OH ↓ 90% (2.76)

product, show that reasonable stereoselectivity can be achieved. A process that has proved particularly useful is to employ as the unsaturated carbonyl compound the enol of a β-diketone or β-ketoester; this results in spontaneous ring-opening of the initial four-membered ring photoproduct by a retro-aldol reaction, and it can give rise to products with medium-sized rings (2.76).

In a different, biological context the photochemical cyclodimerization of carbonyl-conjugated alkenes is important as a major source of ultraviolet-induced damage to living cells. Thymine (2.77)

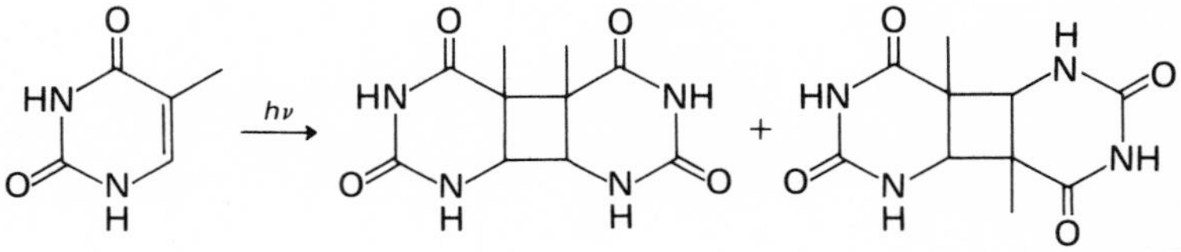

(2.77)

and related bases in nucleic acids undergo dimerization on irradiation, and similar reactions occur between adjacent bases in DNA itself. Although there is a repair mechanism for damaged sections of the DNA sequence, the ultraviolet irradiation of cells can result in mutation or death, a phenomenon that forms the basis for ultraviolet sterilization procedures.

There are a number of other mechanisms by which alkenes can undergo photochemical (2 + 2) cycloaddition, one of which works well for electron-rich alkenes and electron-acceptor sensitizers. The pathway is through the radical cation of the alkene, which attacks a second, ground-state alkene molecule and then cyclizes and accepts an electron to give the product cyclobutane. Typical of this group of reactions is the formation of 1,2-dialkoxycyclobutanes from alkoxyethylenes with dicyanonaphthalene as sensitizer (2.78).

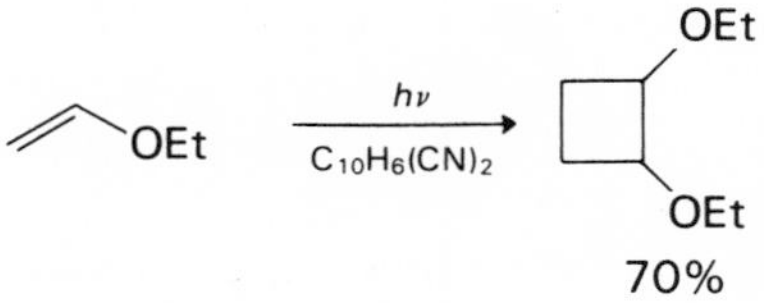

OEt $\xrightarrow[C_{10}H_6(CN)_2]{h\nu}$ OEt, OEt 70% (2.78)

Cyclobutane dimers can be produced from cyclohexenes or cycloheptenes on direct or sensitized irradiation; for cyclohexene itself (2.79) the ratio of isomers is the same in either case, which is

$h\nu$ (p-xylene) 42% + 27% + 23% (2.79)

$h\nu$ $CuOSO_2CF_3$ 57% (2.80)

OH $h\nu$ $CuOSO_2CF_3$ OH ~100% (2.81)

+ OH $h\nu$ $CuOSO_2CF_3$ OH 76% (2.82)

consistent with initial isomerization to a *trans* isomer, followed by attack of this highly strained *trans* species on ground state, *cis*-cyclohexene. Similar reactions are promoted by copper(I) compounds, although in this case metal-complexed species must be involved. Cycloheptene gives largely one stereoisomer of its dimer under these conditions (2.80), and the method is effective for promoting intramolecular cycloaddition in acyclic 1,6-dienes (2.81) and for encouraging cross-products to predominate when a mixture of alkenes is used (2.82).

Photocycloadditions of higher order than (2 + 2) are sometimes encountered, but they are not so general as the (2 +2) reactions. Often they arise in reactions that occur by way of radical cations (2.83), when electrophilic attack on an aromatic ring may divert the reaction from cyclobutane formation, or in those that are promoted

Ph Ph NC CO2Me hν Ph Ph Ph Ph Ph Ph Ph 70% (2.83)

hν CuOSO2CF3 + 80% overall (2.84)

by metal species (2.84), in which case the controlling factors may be the relative orientation of ligands co-ordinated to the metal atom.

Photo-oxidation

Molecular oxygen makes frequent appearances in accounts of organic photochemistry, often in the role of a nuisance. It is a reactive compound and combines readily with radical species that are intermediates in many photochemical reactions, which may result in the appearance of unwanted by-products. Photo-oxidation by this mechanism plays a large part in the degradation of hydrocarbon polymers, leading to loss of mechanical strength and discoloration on exposure to light. In principle, oxygen in its ground state (which is a triplet) could react more rapidly with compounds in their excited state than in their ground state, but in practice few photo-oxidations occur by this route. Much more common is the interaction by energy transfer from substrate excited state (triplet) to ground state oxygen, giving singlet excited oxygen (2.85), or by electron transfer from substrate excited state (singlet or triplet) to ground state oxygen, giving radical ions (2.86). The occurrence of these rapid reactions means that small amounts of oxygen can quench a photophysical or photochemical process efficiently, and so lead, for example, to

$$M(T_1) + O_2(T_0) \longrightarrow M(S_0) + O_2(S_1) \qquad (2.85)$$

$$M(S_1 \text{ or } T_1) + O_2(T_0) \longrightarrow M^{\cdot +} + O_2^{\cdot -} \qquad (2.86)$$

inaccurate measurements of quantum yields or lifetimes, as well as providing further mechanisms for the formation of oxidized products.

Singlet excited oxygen (represented as $^1O_2$) can be generated using an added low-energy triplet sensitizer, so that compounds such as alkenes, which do not readily form their own triplet state by intersystem crossing after excitation, can be conveniently oxidised. Coloured dyes such as methylene blue (2.87) or Rose Bengal (2.88) are commonly used as sensitizers, because they absorb visible light and do not require the use of ultraviolet radiation which may be absorbed by the substrates or their photo-oxidation products. Singlet oxygen reacts with alkenes in a number of different ways. Simple alkenes containing allylic hydrogen may react to form allylic hydroperoxides (2.89). The process does not go through discrete radical intermediates formed by hydrogen abstraction, since a shift of the double bond occurs invariably (2.90); it can be represented as a concerted process

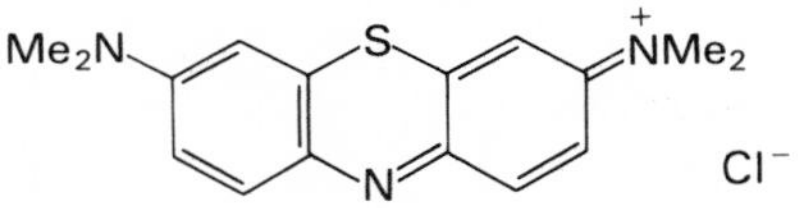

(2.87)

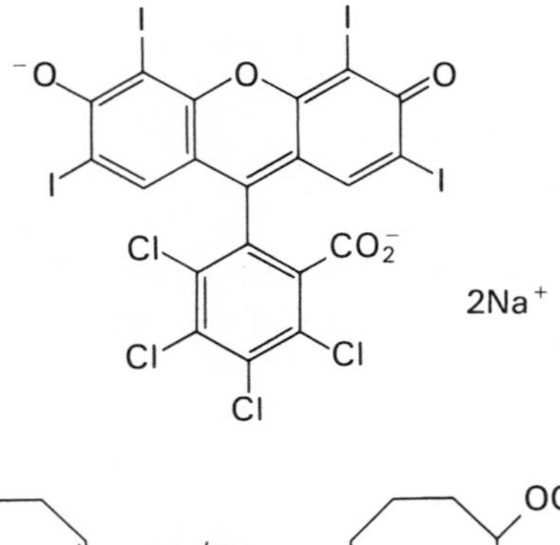

(2.88)

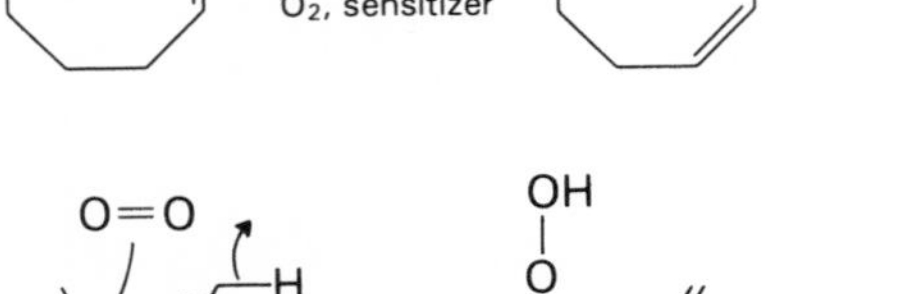

(2.89)

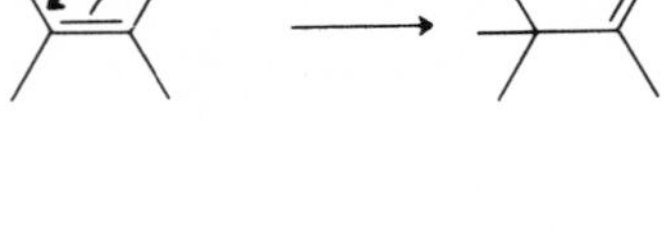

(2.90)

as shown, though a more complex mechanism seems likely. Alkenes without allylic hydrogen, or those with electron-rich double bonds (2.91), undergo a cycloaddition with singlet oxygen to form a 1,2-dioxetane. These can be remarkably stable compounds, bearing in mind that they are strained peroxides, although in some cases they cannot be isolated but break down to give two carbonyl compounds (2.92).

MeO OMe —hν, O_2, sensitizer→ MeO OMe O—O 45–50% (2.91)

—hν, O_2, sensitizer→ O O → O O (2.92)

A (4 + 2) cycloaddition can take place between singlet oxygen and conjugated dienes that are free (or constrained) to adopt an *s-cis* conformation (2.93). When such a reaction occurs with an aryl-conjugated alkene (2.94), the first-formed cycloadduct is itself a conjugated triene, and the overall reaction produces a bis-(4 + 2) adduct.

Ph —hν, O_2, sensitizer→ Ph O O 100% (2.93)

—hν, O_2, sensitizer→ O O → O–O O O 30% (2.94)

Additions of singlet oxygen can be represented as concerted reactions, but extensive work points to the operation of several mechanistic pathways, and the balance between them is influenced by such factors as solvent polarity, as well as by the structure of the substrate. A type of intermediate that has received frequent mention in the debate about mechanisms is a 'perepoxide' (or 'peroxirane', 2.95), but conclusive evidence for its existence is elusive.

$$\text{alkene} \xrightarrow{^1O_2} \text{perepoxide} \longrightarrow \text{products} \qquad (2.95)$$

Whatever the outcome of the mechanistic debate, the reactions of singlet oxygen with alkenes can be very useful in synthesis for producing oxygenated compounds of different types from unsaturated hydrocarbon precursors. The reactions that occur under conditions of electron transfer (rather than energy transfer) have not been so widely investigated, but they differ significantly from the reactions with singlet oxygen. This is not surprising, since the superoxide radical anion ($O_2^{\cdot -}$) is a powerful oxidizing agent, and the alkene radical cation (2.86) is expected to react in a different way from the alkene itself. As with the reactions of singlet oxygen, there are different possible mechanisms for photo-oxidation following the initial electron transfer, such as oxidation of ground-state alkene by $O_2^{\cdot -}$, or oxidation of alkene radical cation by ground-state oxygen, or reaction of the alkene radical cation with superoxide radical anion. Typical outcomes are the formation of a dioxetane, as with singlet oxygen, or the production of 1,2-dioxanes from 1,1-diarylethylenes (2.96) by a route involving attack of alkene radical cation on a second, ground-state alkene molecule prior to trapping by ground-state oxygen.

$$Ar_2C{=}CH_2 \xrightarrow[O_2,\ C_{14}H_8(CN)_2]{h\nu} \text{3,3,6,6-tetraaryl-1,2-dioxane} \qquad (2.96)$$

$>97\%$ ($Ar = p-MeOC_6H_4$)

Alkynes

Alkynes (acetylenes) are like alkenes in that their electronic structure can be described fairly simply in terms of only σ and π bonding and anti-bonding orbitals, but they are different in that the vertical (π,π*) electronic states of alkynes relax to states of fairly rigid *transoid* geometry. Alkynes also have a tendency to produce polymeric material on irradiation, which detracts from the synthetic value of their photoreactions and complicates mechanistic studies. The result of this is that there are not many alkyne photoreactions that correspond to the interesting and useful isomerizations, electrocyclic reactions, sigmatropic shifts, and di-π-methane processes for alkenes, although there are a few exceptions to this generalization, such as the formation of a cyclopropene from tetraphenylpropyne (2.97) in a reaction that parallels the di-π-methane rearrangements described for alkenes.

Although the excited states of alkynes are known to be attacked very rapidly by proton acids, the products obtained by photoaddition of alcohols (2.98) often arise by a radical mechanism that results in hydrogen abstraction from the position adjacent to the alcohol hydroxyl group, rather than by an ionic mechanism, which would result in abstraction of the −OH proton.

By comparison with these less interesting reactions, photocycloadditions of alkynes to alkenes and related compounds provide a fascinating extension of the analogous reactions for alkenes. The simplest type of cycloaddition yields a cyclobutene (2.99), but depending on the relative absorption characteristics of the substrates and product at the wavelength of irradiation, the cyclobutene may

$$Ph—C\equiv C—CPh_3 \xrightarrow{h\nu} \text{(tetraphenylcyclopropene) } 46\% \quad (2.97)$$

$$\text{1-hexyne} + \text{2-propanol (OH)} \xrightarrow{h\nu} \text{HO-substituted alkene product} \quad (2.98)$$

$$Ph—C\equiv C—Ph + \text{cyclohexene} \xrightarrow{h\nu} \text{(diphenylbicyclo[4.2.0]octene) } 43\% \quad (2.99)$$

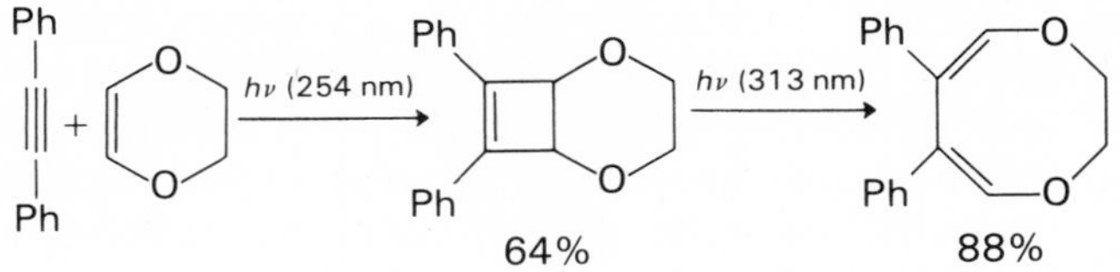

Ph
Ph
hν (254 nm)
Ph
Ph
64%
hν (313 nm)
Ph
Ph
88%

(2.100)

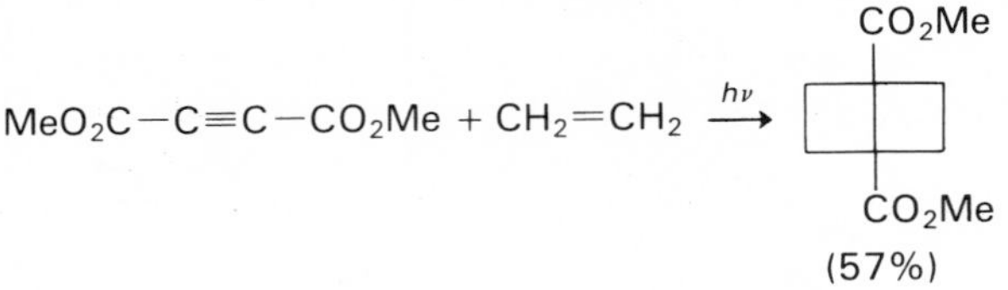

MeO2C—C≡C—CO2Me + CH2=CH2
hν
CO2Me
CO2Me
(57%)

(2.101)

undergo electrocyclic ring-opening (2.100) or a second photocycloaddition with further alkene (2.101).

The reactions with α,β-unsaturated carbonyl compounds also lead to cyclobutenes (2.102), and there is evidence that, in some cases at least, the mechanism is non-concerted and goes through biradical intermediates that can be trapped by a second molecule of the conjugated alkene (2.103). Intramolecular photocycloadditions offer routes to polycyclic structures, and the cyclobutene unit in the product provides a basis for subsequent chemical transformations such as oxidation (2.104).

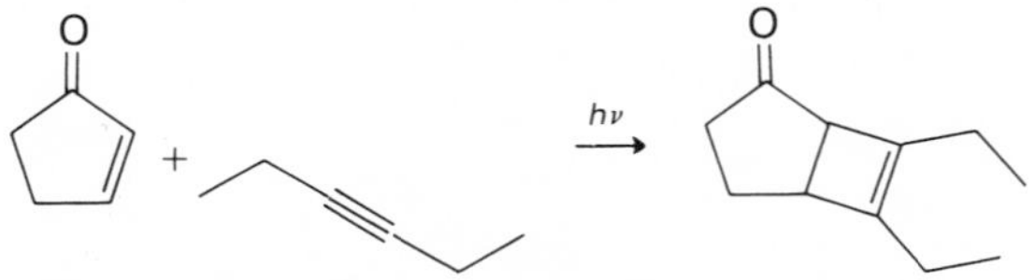

hν

(2.102)

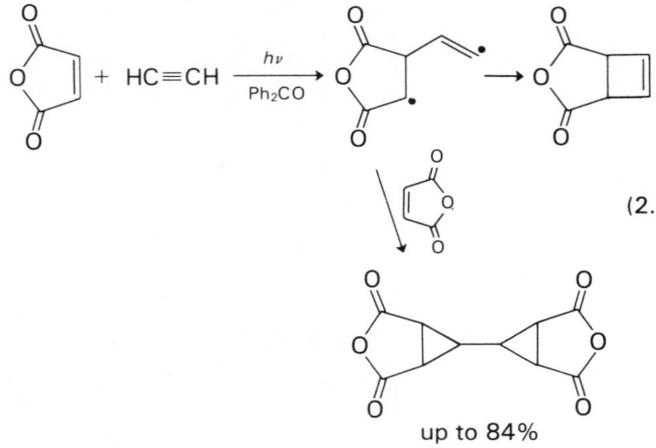

+ HC≡CH
hν
Ph2CO
up to 84%

(2.103)

Alkynes do not dimerize photochemically to give cyclobutadienes, but dimers are formed from arylalkynes under conditions of electron-transfer sensitization (2.105). These dimers arise from a reaction of the alkyne radical cation with ground-state alkyne, followed by intramolecular electrophilic attack on the benzene ring.

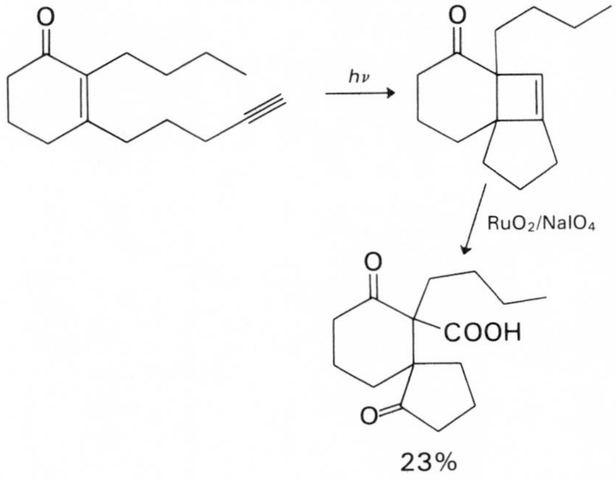

$h\nu$; $RuO_2/NaIO_4$; COOH ; 23% (2.104)

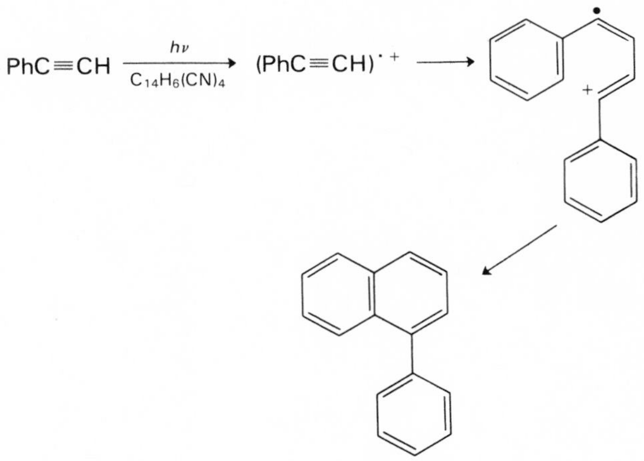

$$PhC{\equiv}CH \xrightarrow[C_{14}H_6(CN)_4]{h\nu} (PhC{\equiv}CH)^{\cdot +} \longrightarrow \quad (2.105)$$

Further reading

P. J. Kropp, in A. Padwa (ed.), *Organic Photochemistry*, vol. 4, Dekker (1979). The photochemistry of alkenes in solution is reviewed, particularly the various rearrangement reactions of monoalkenes.

J. Saltiel and J. L. Charlton, in P. de Mayo (ed.), *Rearrangements in Ground and Excited States*, vol. 3, Academic Press (1980). This chapter provides a detailed account of photochemical *cis–trans* isomerization in olefins (alkenes).

W. G. Dauben, E. L. McInnis and D. M. Michno, in P. de Mayo (ed.), *Rearrangements in Ground and Excited States*, vol. 3, Academic Press (1980). Photochemical rearrangements of trienes are described here, especially electrocyclic reactions and related processes.

H. E. Zimmerman, in P. de Mayo (ed.), *Rearrangements in Ground and Excited States*, vol. 3, Academic Press (1980). The di-π-methane rearrangement is reviewed extensively and its mechanism discussed.

A. C. Weedon, in W. M. Horspool (ed.), *Synthetic Organic Photochemistry*, Plenum (1984). This review covers the use of enone photochemical cycloadditions in organic synthesis, with extensive tabulated data and many synthetic strategies.

A. A. Frimer, *Chemical Reviews*, volume 79, American Chemical Society (1979), p. 359. The reactions and their mechanisms of alkenes with singlet oxygen are the subjects of this review.

S. L. Mattes and S. Farid, in A. Padwa (ed.), *Organic Photochemistry*, vol. 6, Dekker (1983). The photochemical electron transfer reactions of olefins are described in this account, which provides an interesting perspective that cuts across classifications based on an overall reaction type.

J. D. Coyle, in A. Padwa (ed.), *Organic Photochemistry, vol. 7* Dekker (1985). This is an account of the photochemistry of acetylenes (alkynes).

CHAPTER 3

Photochemistry of aromatic compounds

Aromatic compounds have a special place in ground-state chemistry because of their enhanced thermodynamic stability, which is associated with the presence of a closed shell of $(4n + 2)$ pi-electrons. The thermal chemistry of benzene and related compounds is dominated by substitution reactions, especially electrophilic substitutions, in which the aromatic system is preserved in the overall process. In the photochemistry of aromatic compounds such thermodynamic factors are of secondary importance; the electronically excited state is sufficiently energetic, and sufficiently different in electron distribution and electron donor–acceptor properties, for pathways to be accessible that lead to products which are not characteristic of ground-state processes. Often these products are thermodynamically unstable (though kinetically stable) with respect to the substrates from which they are formed, or they represent an orientational preference different from the one that predominates thermally.

The major classes of photochemical reaction for aromatic compounds are nucleophilic substitution and a range of processes that lead to non-aromatic products—valence isomerization, addition or cycloaddition reactions, and cyclization involving 6-electron systems. These five general categories of reaction will be described in the following sections, together with a few examples of more specific processes.

Substitution reactions

Simple benzenoid compounds typically undergo substitution reactions in their electronic ground state, most commonly electrophilic substitution. The relative rates of reaction and the preferred positions of substitution are governed by the nature and position of substituents already present in the ring, and a rationalization can be provided in terms of stabilizing effects on ionic intermediates (and hence on the transition states of related structure that lead to these intermediates). In a few instances products are formed under conditions of thermodynamic rather than kinetic control, and an alternative explanation is required for the observed preferences. For excited-state processes in the photochemical substitution reactions of aromatic compounds, this type of rationalization is inappropriate, because the reacting species (the electronically excited state) is higher in energy than likely ionic intermediates, and influences on the stability of intermediates no longer reflect the major effects on transition state energies (and therefore on relative rates of reaction).

The great majority of reported photosubstitution reactions in benzene derivatives involve attack on the ring by nucleophilic species. Simple reagents that take part in such processes may be anions, such as hydroxide (HO^-) or cyanide (NC^-), or neutral nucleophiles, such as ammonia (NH_3) or amines (RNH_2); common leaving groups are halide, methoxide (MeO^-) or nitrite (NO_2^-) anions. As an example, *m*-nitroanisole reacts on irradiation with potassium cyanide in solution to give *m*-nitrobenzonitrile (3.1) by replacement of the methoxy group. However, for the same substrate with ammonia it is the nitro group that is replaced (3.2), and cyanide displaces the fluorine rather than the methoxy group of *p*-fluoroanisole (3.3).

With suitably substituted benzene derivatives a preference can be seen in the product distributions for replacement of a group *meta* to the substituent which is not displaced (and which can be reckoned as

$$m\text{-}MeOC_6H_4NO_2 \xrightarrow[KCN]{h\nu} m\text{-}NCC_6H_4NO_2 \quad (44\%) \qquad (3.1)$$

$$m\text{-}MeOC_6H_4NO_2 \xrightarrow[NH_3]{h\nu} m\text{-}MeOC_6H_4NH_2 \quad (89\%) \qquad (3.2)$$

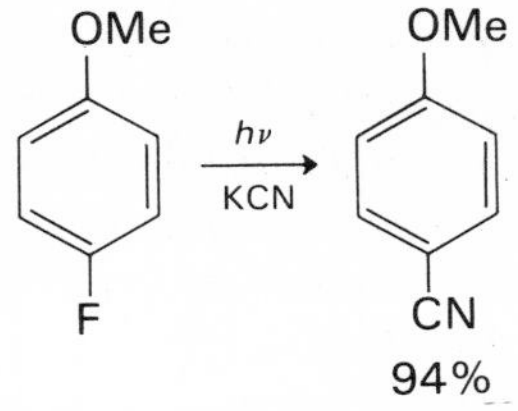

(3.3)

having the greatest influence). So, for 1,2-dimethoxy-4-nitrobenzene, photochemical reaction with sodium hydroxide produces 2-methoxy-5-nitrophenol (3.4), whereas the corresponding thermal reaction leads to 2-methoxy-4-nitrophenol by substitution of the alternative methoxy group. Another example with substituents of different electronic nature is 3,4-dichloroaniline, which with water gives rise to a *meta*-substitution product on irradiation (3.5).

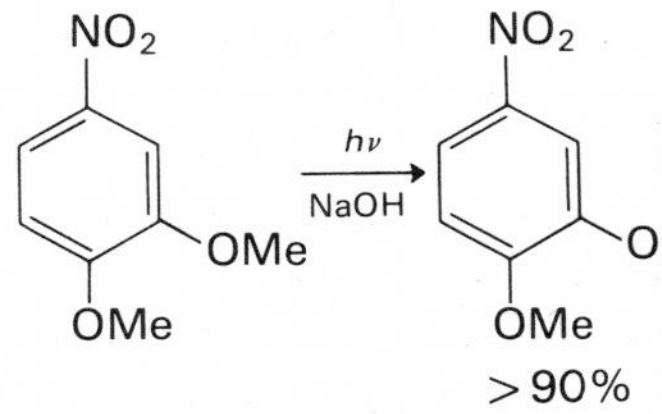

(3.4)

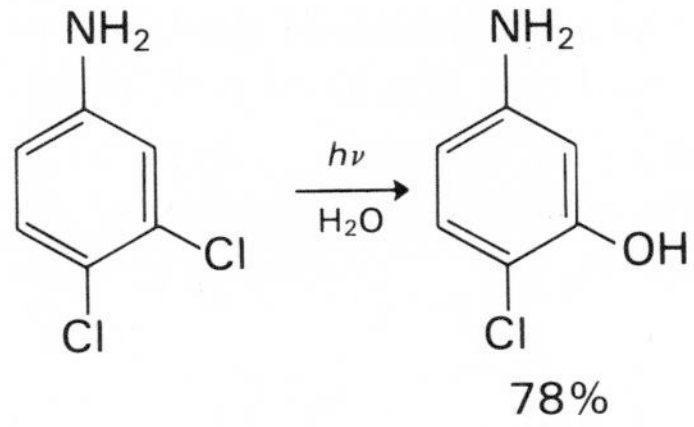

(3.5)

A survey of many such reactions suggests that there is no single, simple pattern that can be used to predict the outcome of photochemical nucleophilic substitutions, but rather a situation in which one of at least three mechanisms may operate, and this has been borne out by more detailed mechanistic studies. One approach to rationalizing the preferred orientation in the excited-state reactions is to calculate electron densities at the various ring carbon atoms for a particular pattern of substituents, and to assume that preferential attack by a nucleophile will take place at the position of lowest electron density. This 'static reactivity' leads to the prediction that a nitro group is *meta*-directing for direct nucleophilic attack in the excited state, because the electron density at this position is substantially lower than in the ground state. (Note that the nitro group is *ortho/para*-directing towards nucleophilic attack in the ground state.) The implication of such a rationalization is that the nucleophile attacks the excited state directly, and it is likely that an intermediate ionic species is produced (3.6).

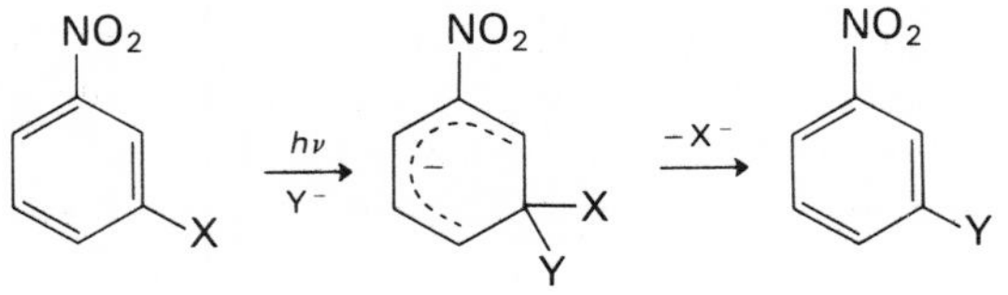

(3.6)

Electron density calculations are less successful in accounting for the reactions of benzenes with substituents such as methoxy, and there is strong evidence with these for a different pathway that involves ejection of an electron to form a radical cation (3.7); this is in keeping with the greatly enhanced electron-donor properties of an excited state. Flash photolysis studies support the formation of radical cations for methoxybenzenes on irradiation, and solvated electrons have also been detected in scavenging experiments. Subsequent attack by the nucleophile on the radical cation can then be rationalized by calculations based on this species rather than on the excited state.

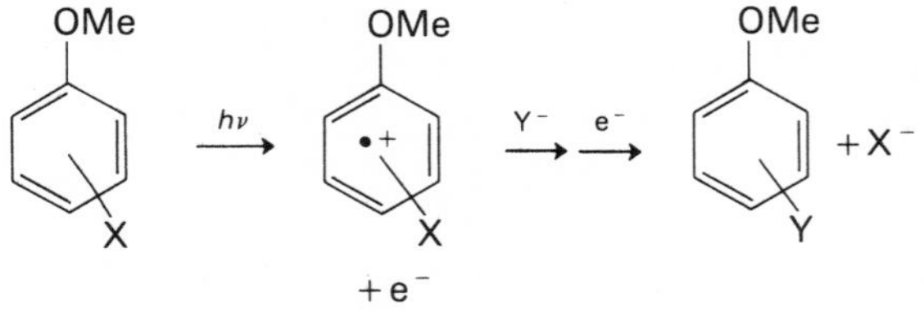

(3.7)

In aromatic radical cations the methoxy group is generally *ortho/para*-directing with respect to an incoming nucleophile. In retrospect the confusion caused by some of the early results may be attributed to the fact that the substrates were benzenes carrying both nitro and methoxy substituents, and it is possible that different mechanisms predominate for systems that appear quite similar. The fine balance that sometimes exists is illustrated by the substrate in reaction (3.4), which undergoes replacement of the *meta*-methoxy group with hydroxide ion or with methylamine (3.8), but gives a *para* substitution product when dimethylamine is employed (3.9), which is a better electron-donor than methylamine.

NO_2 OMe OMe $\xrightarrow[MeNH_2]{h\nu}$ NO_2 NHMe OMe 81% (3.8)

NO_2 OMe OMe $\xrightarrow[Me_2NH]{h\nu}$ NO_2 OMe NMe_2 68% (3.9)

Radical cations can be formed by irradiation of unsubstituted aromatic hydrocarbons such as naphthalene, and this makes possible the photochemical displacement of hydride ion by a nucleophile such as cyanide (3.10). Oxygen is not necessary for the success of this type of reaction if a good electron-acceptor is present, such as *p*-dicyanobenzene (3.11), which enhances the initial photo-ionization and also provides for reaction with the displaced hydrogen.

$\xrightarrow[KCN,\ O_2]{h\nu}$ CN 65% (3.10)

OMe $\xrightarrow[KCN,\ C_6H_4(CN)_2]{h\nu}$ OMe CN + OMe CN 61% (1:1 ratio) (3.11)

The synthetic utility of many of the substitution reactions described so far is limited because there are well-established thermal routes to the same products. However, a third group of photochemical nucleophilic substitutions involves aryl halides and nucleophiles based on sulfur, phosphorus or, of particular importance, carbon. Two examples are the reaction of bromobenzene with the anion of *t*-butyl methyl ketone (3.12), and the replacement of bromine by cyanomethyl in 2-bromopyridine (3.13). This type of reaction offers a clear advantage over lengthy thermal alternatives, and intramolecular versions have been used in the synthesis of indoles (e.g. 3.14) or benzofurans from *o*-iodoaniline or *o*-iodoanisole respectively.

Br O + $\xrightarrow[KOBu^t]{h\nu}$ O 90% (3.12)

+ KCH_2CN $\xrightarrow{h\nu}$ N Br N CN 75% (3.13)

I NH$_2$ + O $\xrightarrow[KOBu^t]{h\nu}$ O NH$_2$ → N H 83% (3.14)

Mechanistic information from these reactions points to the initial formation of a radical *anion* of the aromatic compound, followed by loss of halide ion (3.15); subsequent attack by a second enolate anion and electron transfer to a second molecule of aryl halide provides the substitution product, and the reaction is propagated. The operation of a chain mechanism is indicated by the observation that quantum yields can be very much greater than unity. The operation of such a radical anion mechanism for photochemical nucleophilic substitution in aromatic compounds, in the light of the radical cation mechanism discussed earlier, highlights the fact that the electron-donor and the electron-acceptor properties of electronically excited states are both enhanced significantly compared with the ground state.

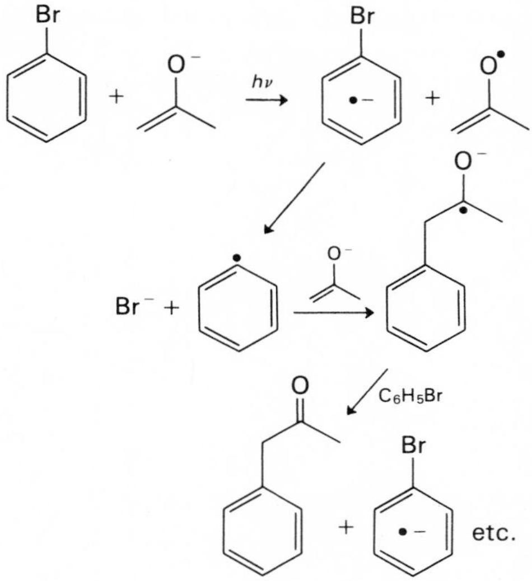

(3.15)

The halogen of aryl halides can be replaced by deuterium on irradiation in the presence of a deuterium atom source such as deuterated methanol (3.16), and this is a very effective way of making specifically deuterated aromatic compounds. A radical-anion mechanism is feasible for such a reaction, although direct homolytic cleavage of the carbon–halogen bond is considered to be more likely. In some contexts replacement of aromatic halogen by hydrogen is desirable for degradative purposes, and in particular the

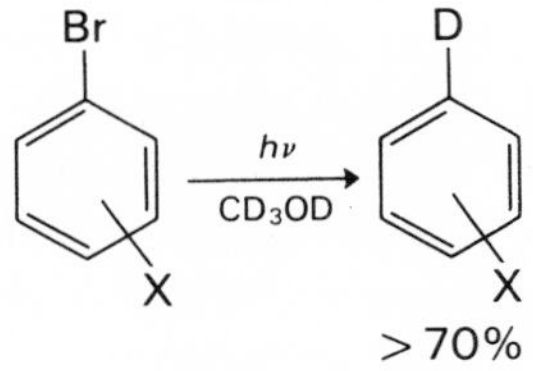

(3.16)

conversion of chlorinated dibenzodioxins (e.g. 3.17) that can arise as by-products in the production of chlorinated phenols for weedkillers such as 2,4,5-T, to less heavily substituted (and less toxic) compounds. Photochemical replacement of chlorine by hydrogen has been carried out to destroy these substances in samples of industrial waste. In the context of synthesis rather than degradation, the generation of aryl radicals by photochemical cleavage of carbon–iodine bond is a very effective way of replacing halogen by phenyl if the reaction is carried out in benzene (e.g. 3.18).

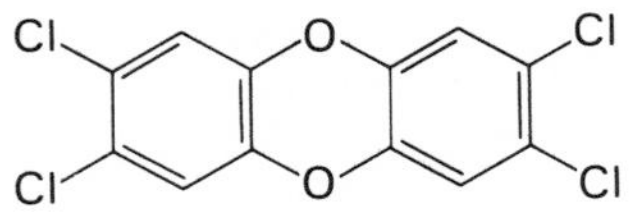

(3.17)

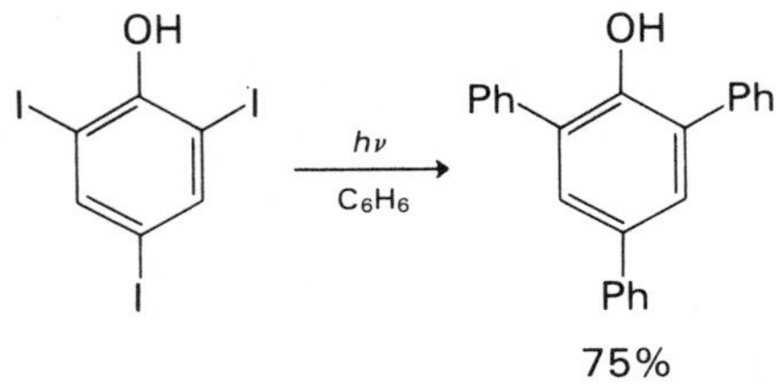

$\xrightarrow[C_6H_6]{h\nu}$ (3.18)

75%

The dearth of information about photochemical electrophilic substitution arises in part because many typical electrophiles are good physical quenchers of excited states; that is they deactivate the excited state without leading to chemical reaction. Studies involving hydrogen isotope exchange suggest that the directing effects of substituents differ from those in ground-state reactions (−OMe is *ortho*-directing, and $-NO_2$ is *para*-directing, in the excited state), but there are too few data for generalizations to be made.

Radical substitution plays a part in the thermal chemistry of aromatic compounds, but not in the photochemistry, except in so far as many radicals that attack aromatic compounds are generated by photochemical methods from other addends. The reason for this is that reactive radicals exist only in low concentrations, and electronically excited states similarly are formed only in low concentrations; the rate of bimolecular reaction between two such reactive species is generally much lower than the rates of alternative processes such as attack of the radical on ground-state aromatic compounds.

A photochemical substitution reaction that does involve radical intermediates, but in a different way, is the photo-Fries reaction (named after its thermal counterpart, the Fries reaction). Typically, irradiation of a phenyl ester of a carboxylic acid leads to homolytic cleavage of the acyl–oxygen bond, and the resultant radicals recombine by formation of acyl to ring-carbon bonds. This occurs preferentially in the *ortho* or *para* positions, and the cyclohexadienones so formed usually tautomerize to produce substituted phenols (3.19). In some systems the photochemical reaction proceeds with less extensive side-reactions than does the thermal analogue, as with esters of 1-naphthol (3.20).

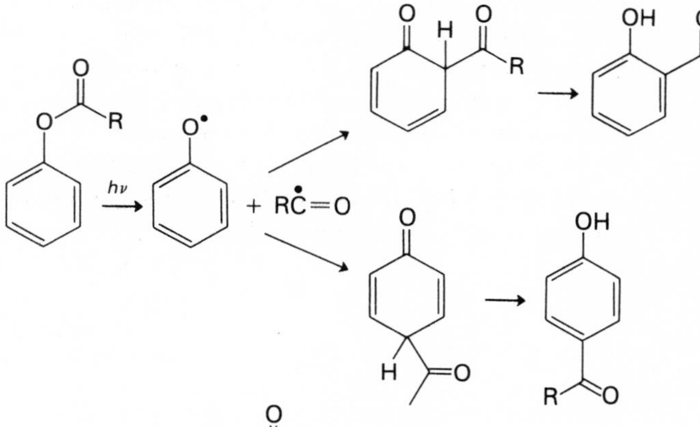

(3.19)

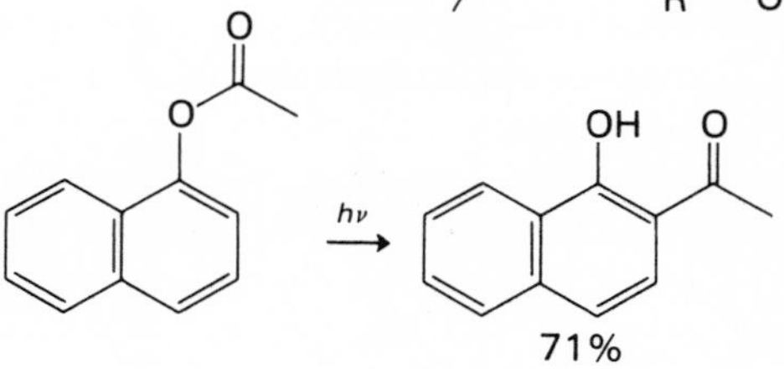

71%

(3.20)

87%

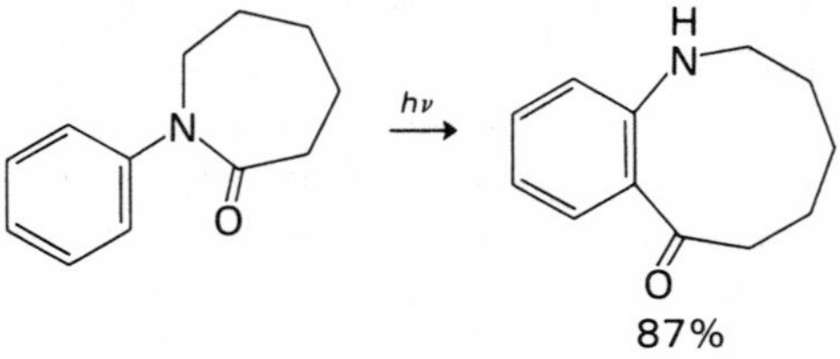

(3.21)

There are a number of reactions related to the photo-Fries process, in which cleavage of a bond adjacent to a heteroatom ring substituent leads by way of radical intermediates to a ring-substituted product. As an illustration, *N*-phenylcaprolactam, which can be regarded as an acylaniline, undergoes efficient photoreaction to give a benzazoninone, which has a nine-membered fused ring system (3.21).

Ring isomerization

When irradiated without added substances many aromatic compounds, especially those carrying alkyl, fluoro or perfluoroalkyl groups, undergo photoisomerization. Sometimes this exhibits itself as a change in the pattern of substitution, as with the xylenes (3.22), and studies of substrates with isotopically labelled ring atoms reveal that a remarkable transposition of adjacent ring carbons, together with their substituents, has occurred. Rationalization of this skeletal transformation invokes the intermediacy of 'valence isomers' of the aromatic system, which are polycyclic and not aromatic. The necessary bonding changes can be accommodated in a two-step mechanism (3.23) involving the formation and subsequent breakdown of such an isomer called a benzvalene.

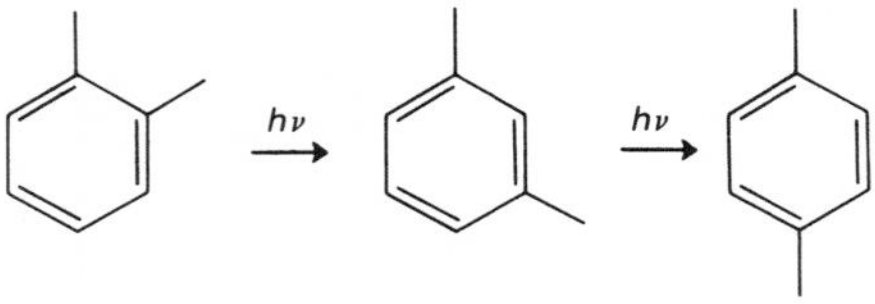

$h\nu$ $h\nu$ (3.22)

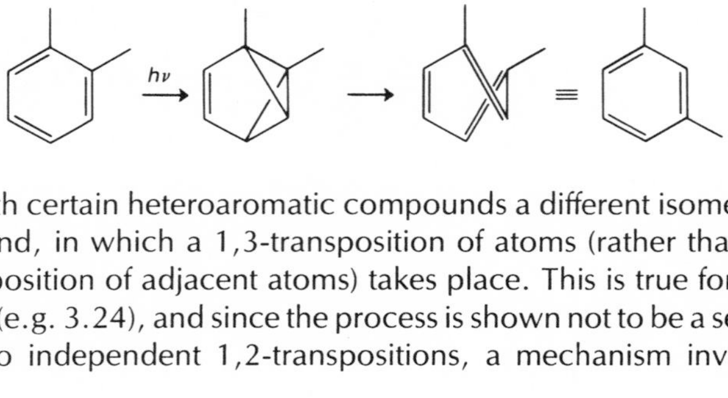

$h\nu$ ≡ (3.23)

With certain heteroaromatic compounds a different isomerization is found, in which a 1,3-transposition of atoms (rather than a 1,2-transposition of adjacent atoms) takes place. This is true for pyridazines (e.g. 3.24), and since the process is shown not to be a sequence of two independent 1,2-transpositions, a mechanism involving a

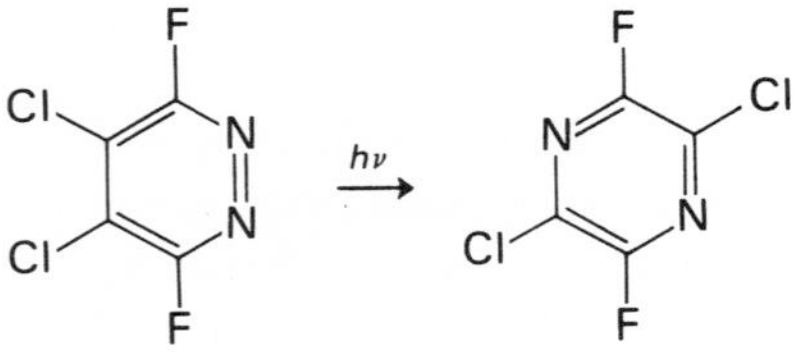

F Cl N N Cl F $h\nu$ F N Cl N Cl F (3.24)

different type of valence isomer is invoked (3.25), namely a diaza derivative of bicyclo[2.2.0]hexadiene.

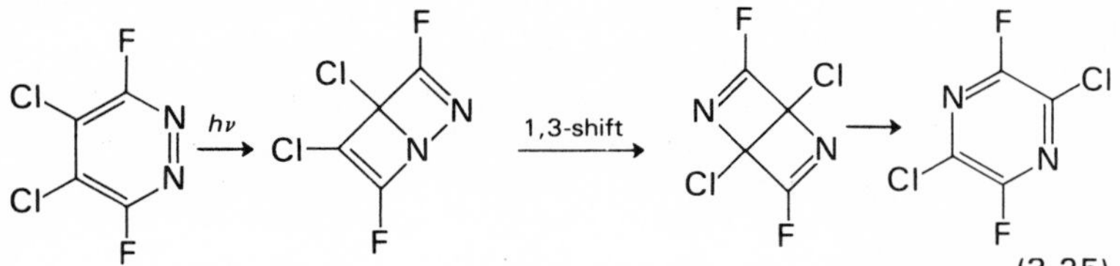

(3.25)

Support for these mechanistic proposals was boosted by the successful isolation of related valence isomers from benzenes or heteroaromatic compounds carrying bulky (*t*-butyl) or fluorinated groups. 1,3,5-Tri-*t*-butylbenzene gives rise to a benzvalene (3.26), but fluorinated substituents or more severe steric crowding (e.g. 3.27) lead to a preference for bicyclo[2.2.0]hexadiene formation (these latter isomers are sometimes called 'Dewar benzenes'), as does the presence of nitrogen in the ring (3.28).

The valence isomers of aromatic compounds can, if they absorb the radiation, undergo further photochemical reaction, and under

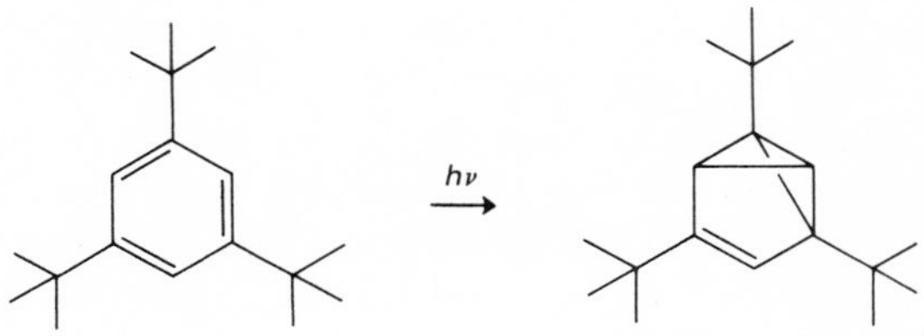

(3.26)

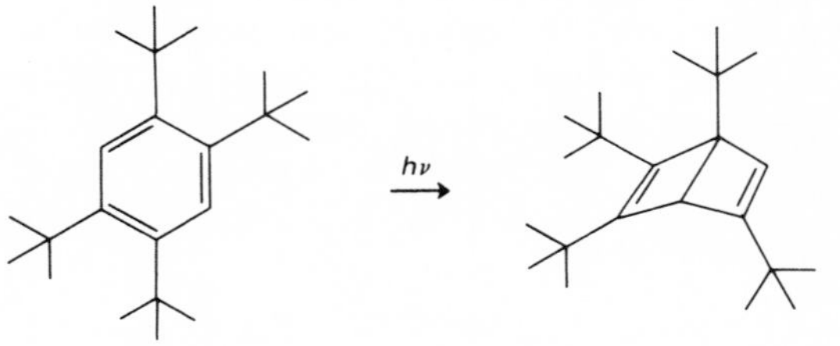

(3.27)

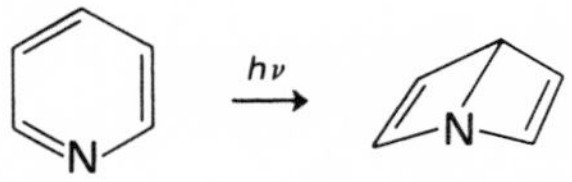

(3.28)

some conditions a photostationary state is achieved between various valence isomers and positional isomers of the original aromatic compound. More commonly, however, a third class of valence isomer, a prismane, is produced by way of an internal photocycloaddition in a bicyclohexadiene (e.g. 3.29), and because the prismane is a saturated compound that does not absorb significantly the radiation normally employed in organic photochemical reactions, it accumulates and becomes the predominant product.

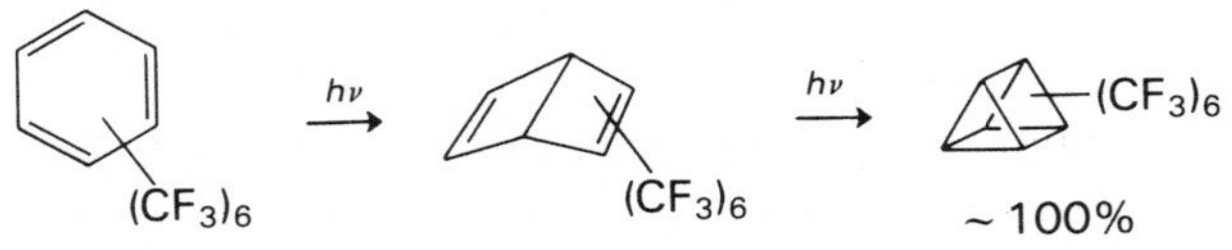

$(CF_3)_6$ $\xrightarrow{h\nu}$ $(CF_3)_6$ $\xrightarrow{h\nu}$ $(CF_3)_6$ ~100% (3.29)

For benzene itself, benzvalene is the major isomer formed in the liquid phase using radiation of wavelength 254 nm (3.30), although it absorbs this radiation and is reconverted efficiently to benzene. With shorter wavelengths bicyclohexadiene and a third isomer, fulvene, are produced (3.31). All of these benzene isomers are very reactive and have been studied only in dilute solutions; it is remarkable that they have even the kinetic ability that they do possess, in the light of the very considerable thermodynamic instability (particularly for benzvalene and bicyclohexadiene) with respect to benzene. In principle, these compounds offer a method for the chemical storage of solar energy, since the release of heat in their reconversion to benzene can be controlled in a catalytic decomposition; however, the major problems remain the need for ultraviolet, rather than visible, radiation to produce the valence isomers from most benzene derivatives, the quantum inefficiencies of the photochemical reactions, and the rather low storage capacity.

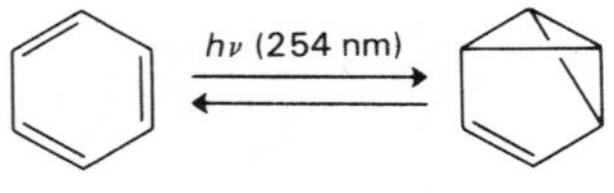

$h\nu$ (254 nm) (3.30)

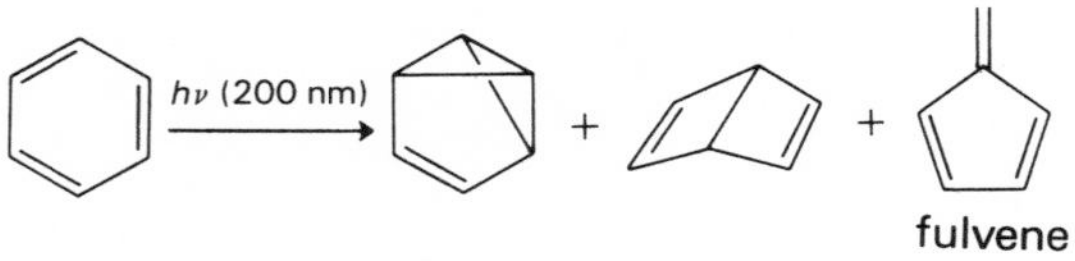

$h\nu$ (200 nm) + + fulvene (3.31)

(3.32)

Although valence isomerization reactions of aromatic compounds have found little by the way of practical application, they are a fascinating area for mechanistic and theoretical study. The details are not completely clear, but it seems that, for benzene itself, benzvalene arises from the lowest excited singlet state, perhaps by way of a biradical intermediate (3.32) that could also be a precursor to fulvene; bicyclohexadiene is probably produced from the second excited singlet state. For some other aromatic compounds the electronic nature of S_1 and S_2 may be reversed, or at least the states are much closer in energy, so that the preference for benzvalene or bicyclohexadiene formation under conditions of long-wavelength irradiation can be rationalized.

Valence isomer formation is a feature also of the photochemistry of naphthalenes (3.33) and anthracenes; for naphthalenes, as for benzenes, the extent of steric crowding helps to determine which type of valence isomer predominates, since there is more severe interaction in the bicyclohexadiene products than in the benzvalene products. Amongst five-membered heteroaromatic compounds there are many known ring photoisomerizations that involve conversion of a 2-substituted to a 3-substituted system (e.g. 3.34). In some cases non-aromatic products can be isolated, such as bicyclo[2.1.0]pentene analogues from thiophenes (3,35), or acylcyclopropenes from furans (3.36); related species may be

R R R R $\xrightarrow{h\nu}$ (3.33)

R R R

R R = C(CH₃)₃ → 94%

CN

$\xrightarrow{h\nu}$ (3.34)

N CN N

H H

55%

F_3C CF_3 F_3C CF_3

$\xrightarrow{h\nu}$ (3.35)

F_3C S CF_3 F_3C S CF_3

intermediates in those ring-transposition reactions where no such compounds are isolated, but the variety of observed isomerizations demands a more complex explanation than one involving only one type of intermediate.

$$\text{tetrakis(trifluoromethyl)furan} \xrightarrow{h\nu} \text{(trifluoroacetyl)tris(trifluoromethyl)cyclopropene} \; (20\%) \qquad (3.36)$$

Addition reactions

Perhaps the best-known photoaddition reaction of benzene is that with chlorine to produce hexachlorocyclohexane (3.37), of which one steroisomer is widely used as an active component in insecticides. However, this reaction does not involve the excited state of benzene; chlorine absorbs light and cleaves homolytically to give chlorine atoms, which then attack the ground state of benzene, leading to overall addition.

$$C_6H_6 + 3Cl_2 \xrightarrow{h\nu} C_6H_6Cl_6 \; (95\%) \qquad (3.37)$$

Photoadditions that arise by initial excitation of the aromatic compound are not common. Benzvalenes are readily attacked by hydroxylic compounds, and so irradiation of benzene in aqueous solutions of acetic acid, for example, results in the formation of a bicyclic product (and an isomer derived from it by subsequent photoisomerization) as a result of addition to the initially formed valence isomer (3.38). A different kind of photoaddition occurs when benzenes react photochemically with amines; cyclohexa-1,4-dienes are the major products (3.39), accompanied by cyclohexa-1,3-dienes, and unlike many of the photochemical reactions of benzene this does not suffer loss of efficiency in scaling-up.

It may seem surprising that one of the most effective reducing hν HOAc AcO hν AcO (3.38)

NH_2 hν H N H N + + 60% (3.39)

agents for the excited states of aromatic compounds is sodium borohydride. Electron-deficient rings (as in benzoate esters) are photoreduced to dihydro and tetrahydro products, and in the presence of an added electron-acceptor hydrocarbons such as naphthalene (3.40), or other electron-rich aromatic compounds, may also be reduced. The mechanistic picture is not yet clear, but it seems evident that radical ions play a major role.

hν $C_6H_4(CN)_2$ (3.40)

$+ NaBH_4$ 59%

Cycloaddition reactions

The irradiation of benzenes with alkenes provides a fascinating array of photochemical reactions, not least because it converts the aromatic substrates into polycyclic, non-aromatic products. In principle, benzene can undergo reaction across the 1,2-(*ortho*), 1,3-(*meta*), or 1,4-(*para*) positions; the 1,3-cycloaddition is structurally the most complex, but it is the predominant mode of reaction for many of the simplest benzene/alkene systems. The products are tricyclic compounds with a fusion of two five-membered rings and one three-membered ring, and an example is the reaction of benzene with vinyl acetate (3.41). For monosubstituted benzenes there can be a high

+ OAc hν (3.41)

OAc $\phi = 0.22$

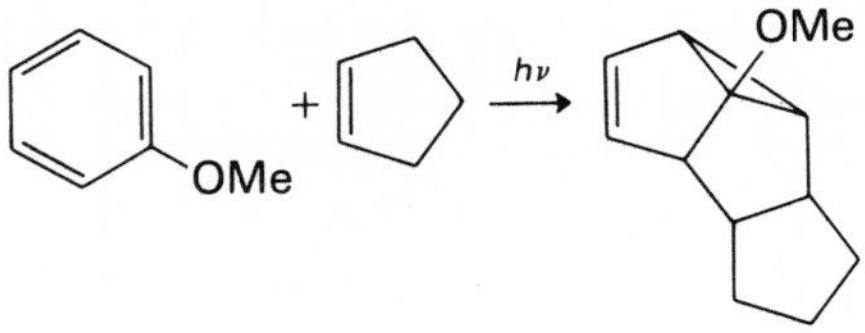

(3.42)

degree of regioselectivity, as seen for the reaction of anisole (methoxybenzene) with cyclopentene (3.42).

A second major mode of photocycloaddition involves 1,2-addition to the aromatic ring, and this predominates if there is a large difference in electron-donor/acceptor capacity between the aromatic compound and the alkene. It is therefore the major reaction pathway when benzene reacts with an electron-rich alkene such as 1,1-dimethoxyethylene (3.43) or with an electron-deficient alkene such as acrylonitrile (3.44). When substituted benzenes are involved, such as anisole with acrylonitrile (3.45), or benzonitrile with vinyl acetate (3.46), reaction can be quite efficient and regioselective to give products in which the two substituents are on adjacent carbon atoms.

One of the first reported photocycloaddition reactions of benzene was with the highly electron-deficient alkene, maleic anhydride.

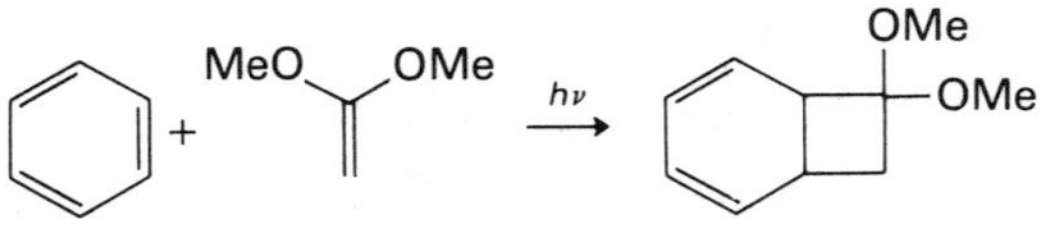

(3.43)

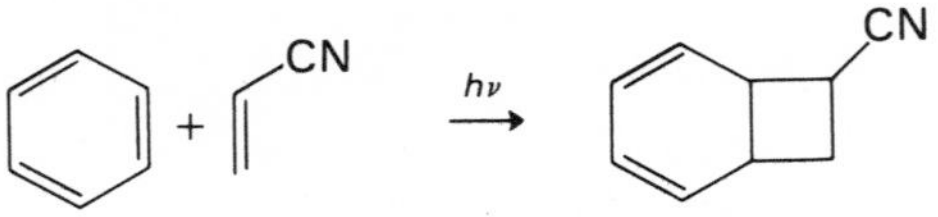

(3.44)

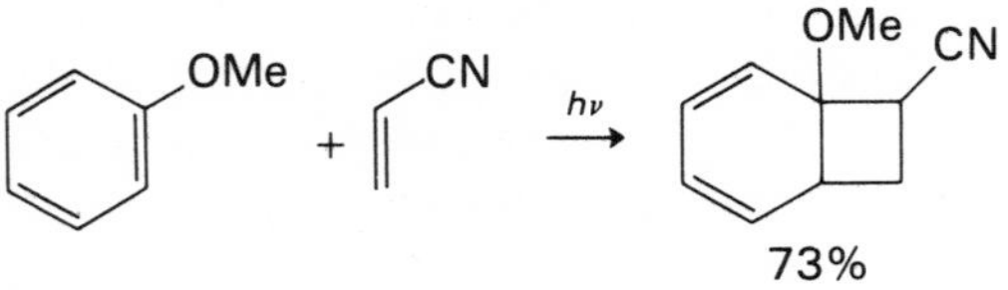

73%

(3.45)

61%

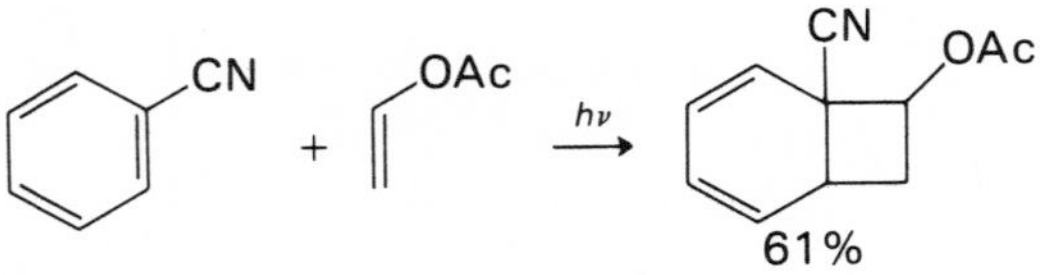

(3.46)

Because the bicyclic product arising from 1,2-cycloaddition contains a cyclohexa-1,3-diene unit, a second molecule of maleic anhydride is incorporated in the isolated product (3.47) as a result of a Diels–Alder reaction with the initial photoproduct. 1,2-Cycloaddition is also the observed mode of reaction with most alkynes, and this leads by way of ring-opening in the initial adduct to a cyclo-octatetraene (3.48).

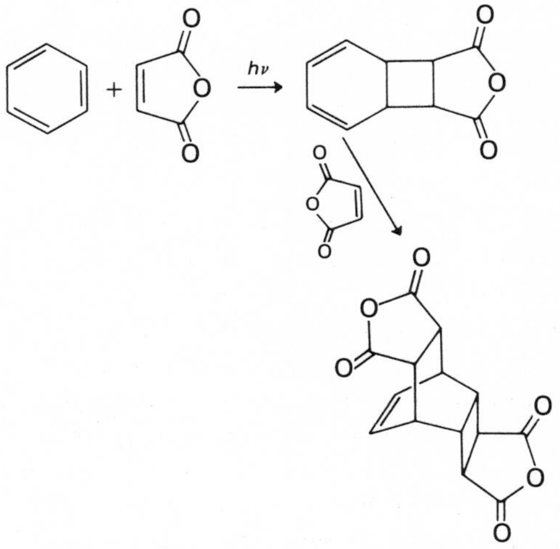

(3.47)

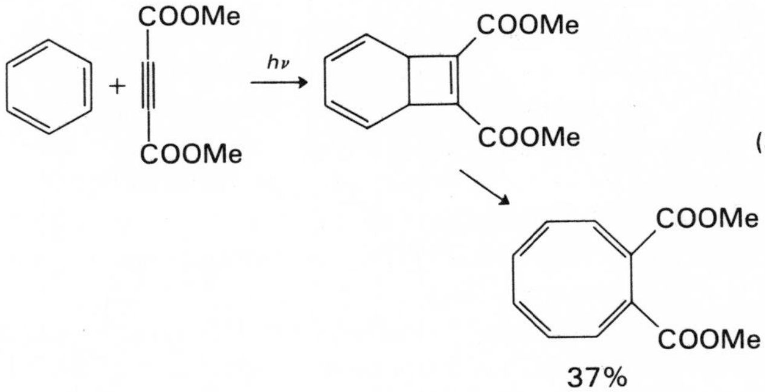

88%

(3.48)

37%

1,4-Cycloaddition is much less frequently encountered in the photoreactions of benzenes with alkenes, but it does provide the major route to product for allenes (3.49). Different 1,4-adducts that involve a (4 + 4), rather than a (2 + 4), cycloaddition accompany the

1,3-adducts obtained when benzene is irradiated with a conjugated diene (3.50); the presence of iodine ensures that the less stable *trans*-cycloadduct isomerizes to the *cis*-compound rather than dimerizing.

$$+ H_2C{=}C{=}CH_2 \xrightarrow{h\nu} \qquad (3.49)$$

$$+ \xrightarrow[I_2]{h\nu} \quad + \qquad (3.50)$$

In most of the reactions described so far in this section the question of stereochemistry arises, though it has not been highlighted in the equations drawn. For all three modes of cycloaddition a particular stereochemical arrangement in the alkene (*cis* or *trans*) is preserved in the adduct, but where new stereochemical features have been introduced (e.g. *exo* or *endo* arrangements in the 1,2 or 1,3-cycloadducts) there is no clear-cut pattern. In some systems there is a strong preference for one type of product, but in others a mixture of stereoisomers is formed. As yet there is not complete certainty about the mechanism of the cycloadditions, and it is likely that there are several different available pathways. The formation of the 1,3-cycloadducts occurs by way of the lowest excited singlet state of the aromatic compound, and it seems to be either a fully concerted reaction, or a concerted addition of the alkene to a pre-formed intermediate (e.g. the prefulvene biradical, 3.32). For 1,2-cycloadditions it is not always the aromatic compound that is excited initially, but rather the alkene or a charge-transfer complex between the aromatic and the alkene, and in these systems there is a greater likelihood that exciplexes, zwitterions or even radical ions play a part in the mechanism.

Synthetic applications of these photocycloadditions to aromatic compounds are sometimes hampered by low chemical yields and poor selectivity in the photoreactions. However, a number of elegant syntheses of tricyclic sesquiterpenes have been based on intramolecular 1,3-photocycloadditions (e.g. 3.51), and these represent a completely new approach to the preparation of such systems.

Condensed aromatic compounds, for example naphthalenes or

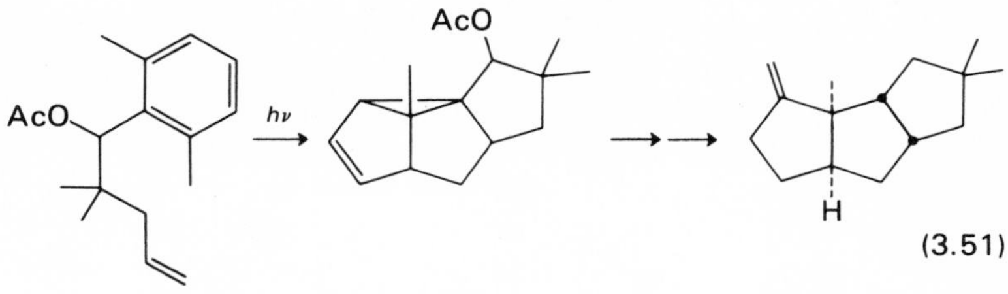

(3.51)

anthracenes, can also be involved in photocycloadditions. Naphthalenes generally give 1,2-cycloadducts across positions 1 and 2 of the naphthalene ring (3.52). The adduct between 2-naphthol and acrylonitrile is obtained with high regioselectivity (3.53), and it can be converted readily to an acyclic adduct that represents overall a formal Michael addition reaction between the initial substrates. The first-formed photoadducts between naphthalenes and alkynes are themselves photolabile, and the product isolated is formed by a subsequent intramolecular photocycloaddition (3.54).

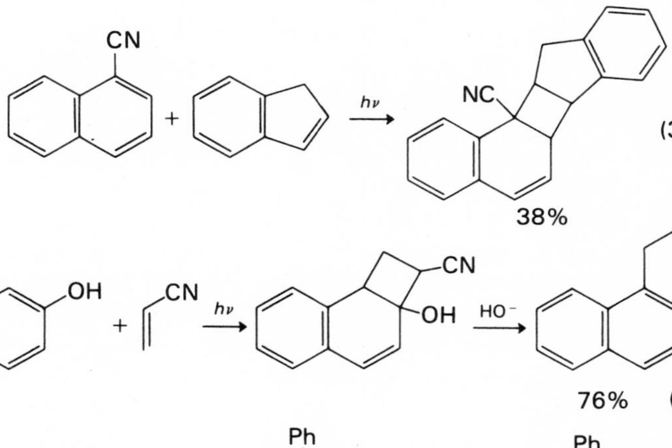

38% (3.52)

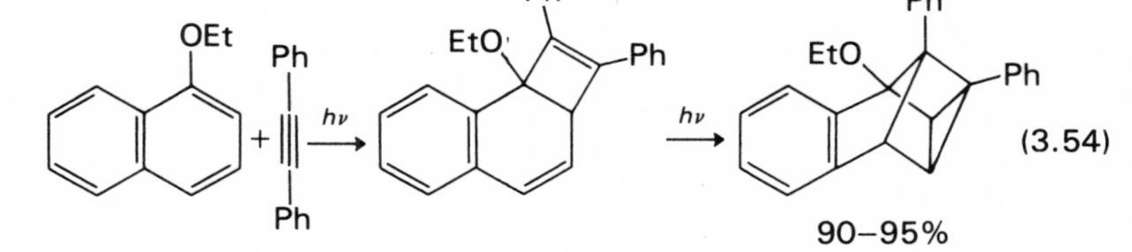

76% (3.53)

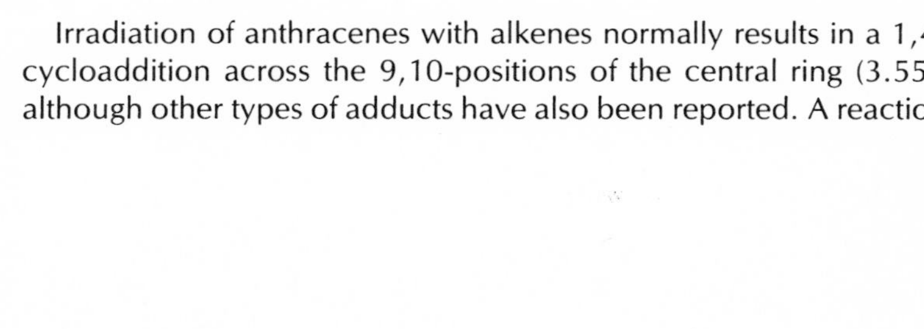

90–95% (3.54)

Irradiation of anthracenes with alkenes normally results in a 1,4-cycloaddition across the 9,10-positions of the central ring (3.55), although other types of adducts have also been reported. A reaction that has been widely studied, and which was one of the earliest reported of all organic photoreactions, is the photodimerization of anthracenes to give products in which 1,4-cycloaddition across the 9,10-positions of both molecules has occurred (3.56). Like the valence isomers of substituted benzenes, the photodimers of substituted anthracenes have been proposed as possible energy-storage compounds for solar radiation; suitable anthracenes can be made to absorb in the visible region, and the stored energy can be recovered in a controlled way by catalysed reconversion of the dimer to monomer, but major drawbacks are the low efficiencies of energy collection and the low storage capacity.

CN + CN $\xrightarrow{h\nu}$ (3.55)

2 $\xrightarrow{h\nu}$ (3.56)

>90%

Heteroaromatic compounds do not undergo the same variety of photocycloadditions with alkenes as do their carbocyclic counterparts. There are very few reports of this type of reaction for six-membered ring compounds such as pyridines, but five-membered ring systems such as furans do give 1,2-cycloadducts with a range of alkenes (e.g. 357).

O + O O O $\xrightarrow[\text{sensitizer}]{h\nu}$ O O O O (3.57)

Aromatic compounds undergo photochemical cycloaddition with a few classes of addend other than those that react by way of a carbon–carbon multiple bond. The most important of these is dioxygen (O_2). On irradiation with oxygen, anthracenes, naphthalenes and similar compounds are converted to cyclic peroxides (e.g. 3.58) as a result of 1,4-addition. The mechanism of the process involves singlet oxygen, which is obtained from the ground-state (triplet) oxygen molecule by triplet energy transfer from an excited triplet state of the aromatic compound. Furans are very good substrates for singlet oxygen cycloaddition (3.59), although in this case an added triplet sensitizer is normally employed to avoid the use of short-wavelength ultraviolet radiation.

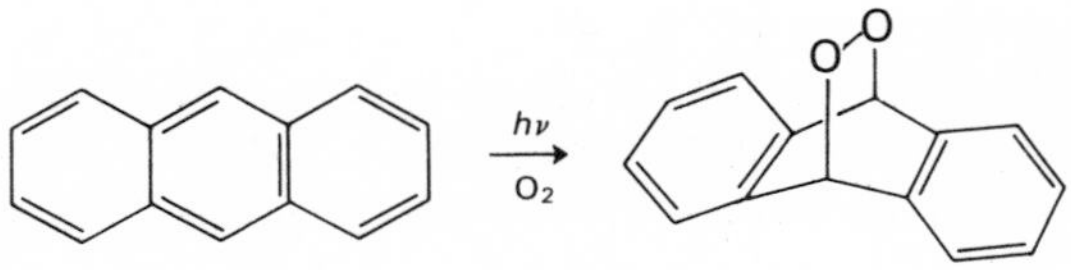

(3.58)

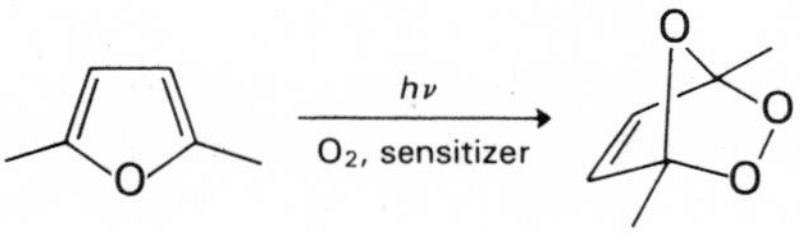

(3.59)

Cyclization reactions

There are a large number of photochemical cyclizations of aromatic compounds that lead initially to polycyclic, non-aromatic products, although subsequent rearrangement, elimination or oxidation occurs in many instances to form aromatic secondary products. The archetype for one major class of photocyclization is the conversion of stilbene to phenanthrene by way of a dihydrophenanthrene (3.60).

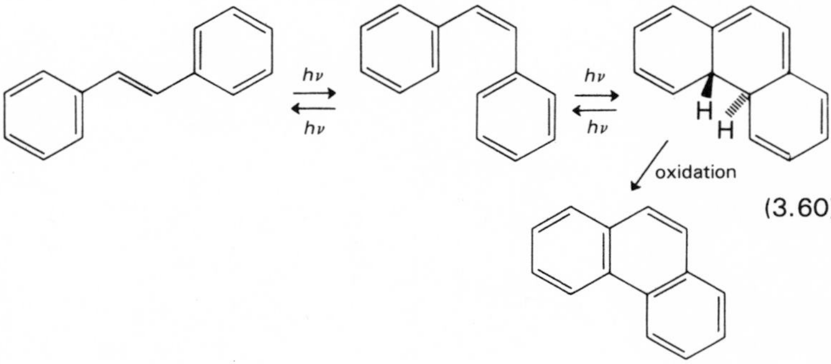

(3.60)

trans-Stilbene is first isomerized photochemically to the *cis*-isomer (a photostationary state is achieved), and in a second, much less (quantum) efficient photochemical reaction *cis*-stilbene cyclizes to 4a,4b-dihydrophenanthrene. This is a 6-electron electrocyclic ring-closure, which occurs in a conrotatory manner (see p. 48) to give the *trans* isomer of the dihydro compound. Under conditions where all oxygen is rigorously excluded the coloured dihydrophenanthrene can reach a moderate photostationary state level, but more commonly it is oxidized to phenanthrene. The most widely used oxidants to enhance this stage are air (oxygen) or iodine.

For some 4a,4b-dihydrocompounds isomerization can occur in protic solvents, so that the isolated product under non-oxidative conditions is a 9,10-dihydrophenanthrene; this is fairly general if the stilbene double bond carries an electron-withdrawing group such as methoxycarbonyl (3.61). The presence of an added primary amine as base can also channel the reaction to give an isolable dihydrophenanthrene, often a 1,4-dihydro derivative (3.62); the function of the amine is probably as a base to isomerize the initially formed 4a,4b-dihydrophenanthrene. Occasionally a phenanthrene product is formed without the need for oxidation, because of elimination from the dihydro compound involving a leaving group that occupied one of the *ortho* positions in the original stilbene (e.g. 3.63).

COOMe $\xrightarrow[\text{MeOH}]{h\nu}$ COOMe $\rightarrow$ COOMe (3.61)

$\xrightarrow[\text{Pr}^i\text{NH}_2]{h\nu}$ 70% (3.62)

OMe $\xrightarrow{h\nu}$ 58% (3.63)

A very large number of substituted phenanthrenes have been made from stilbenes by this photocyclization method, as have more complex polycyclic aromatic compounds by related reactions involving a single cyclization (e.g. 3.64 for chrysenes) or two, or more, successive cyclizations (e.g. 3.65). The reaction can be nicely adapted to provide a route from 1-benzylidenetetrahydroisoquinolines to alkaloids of the aprophine family (e.g. 3.66).

NC $\xrightarrow{h\nu}$ NC 76% (3.64)

$\xrightarrow[I_2]{h\nu}$ (3.65)

MeO MeO N COOEt $\xrightarrow[I_2]{h\nu}$ MeO MeO N COOEt 35% (3.66)

Heterocyclic variations on the basic stilbene → phenanthrene theme are provided by many systems in which one or both of the phenyl groups of stilbene is replaced by a heteroaromatic group, or in which there is a pair of phenyl groups on adjacent positions in a heterocyclic ring. An example of the latter is seen in 1,2,6-triphenylpyridinium salts, which undergo photochemical cyclization to give a bridgehead aza-derivative of triphenylene (3.67).

The 6-electron system required for electrocyclic ring-closure can be made up from any combination of aromatic and alkene double

(3.67)

85%

bonds. In stilbene the arrangement is aromatic–alkene–aromatic, but the reaction works well for *o*-vinylbiphenyl (3.68) with an aromatic–aromatic–alkene sequence of bonds, and for compounds such as 2,3-di(benzylidene)butyrolactone (3.69) in which the reactive sequence is aromatic–alkene–alkene. In both of these reactions the isolated product is derived from the initial one, not by oxidation but rather by a 1,5-shift or a 1,3-shift of hydrogen. When such a hydrogen shift is prevented by the use of methyl substitution the initial product can be isolated, and this has been employed in the development of a chemical actinometer (3.70) in which the forward reaction proceeds with ultraviolet light to give a highly coloured product that can be estimated spectrophotometrically; the actinometer is regenerated by reversing the reaction using visible light, and the material can be used many times with no observable 'fatigue' (i.e. no side-reactions).

(3.68)

100%

Ph Ph Ph

100% (3.69)

$h\nu$ (366 nm)

$h\nu$ (500 nm)

(3.70)

100%

Further extensions of the stilbene photocyclization are seen in analogous reactions of compounds containing the imine chromophore (e.g. 3,71) or an amide group (3.72). The amide reaction can be considered formally as giving a zwitterion intermediate, which undergoes proton shifts and oxidation to form the observed product. Non-oxidative cyclizations that start with either *N*-vinyl aromatic carboxamides (C=C−N−CO−Ar) or *N*-aryl α,β-unsaturated carboxamides (Ar−N−CO−C=C) have been extensively used to make quinoline or isoquinoline alkaloids and their derivatives; a fairly simple example is given in (3.73).

OMe hν OMe 56% (3.71)

O NR hν O⁻ NR oxidation O NR (3.72)

hν N H O N H O 82% (3.73)

Diaryl amines, ethers or sulphides, or their aryl vinyl analogues, provide another 6-electron system related to stilbene, but for these a pair of electrons is provided by the single heteroatom (ArXAr or ArXC=C). With diaryl compounds the initial photocyclized product is a zwitterion that undergoes a proton shift to give, for example, *N*-methyl-4a,4b-dihydrocarbazole from methyldiphenylamine, with subsequent oxidation to *N*-methylcarbazole (3.74). With the aryl vinyl analogues the product after the proton shift can normally be isolated (3.75). An especially useful variation of this reaction employs a substrate derived from a β-diketone or β-ketoester; for example, the compound derived from *N*-methylaniline and ethyl acetoacetate reacts very efficiently to give an indole (3.76).

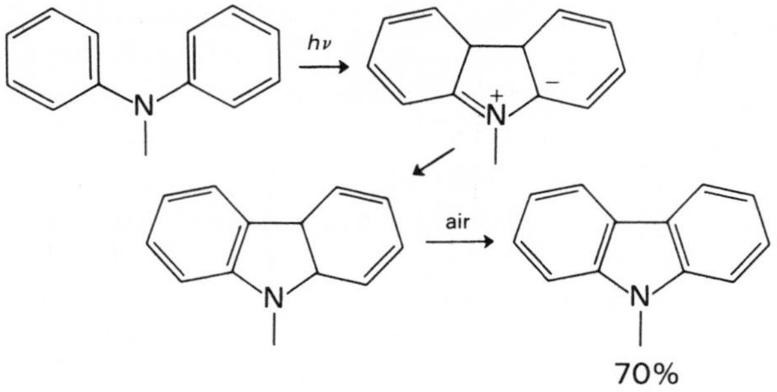

$h\nu$ air 70% (3.74)

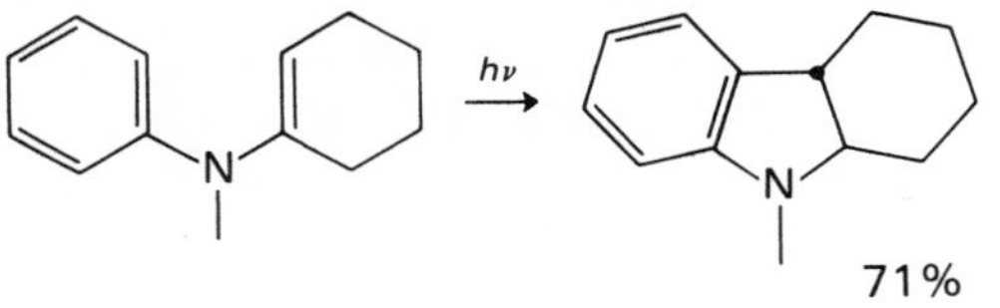

$h\nu$ 71% (3.75)

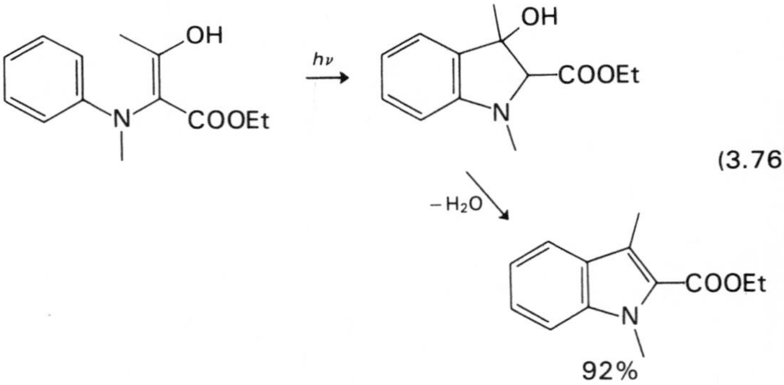

OH COOEt $h\nu$ OH COOEt $-H_2O$ COOEt 92% (3.76)

In the remainder of this section are described some photocyclizations that are different from the stilbene-like processes discussed so far, in that there is no possibility of a 6-electron electrocyclic mechanism because of a break in the conjugation of the system. Some of these are initiated by photochemical cleavage of an aryl–halogen bond to yield an aryl radical that can attack another aromatic ring within the molecule; such a process works best with the iodo-derivatives (e.g. 3.77). An apparently similar cleavage in various chloroacetamido compounds containing a suitably placed aromatic group has been used to make medium-ring fused nitrogen heterocycles (e.g. 3.78), including some complex multicyclic alkaloids; however, the mechanism is different in that electron transfer from the aromatic ring to the chloro-acetamido group, followed by loss of chloride ion and attack by the aliphatic radical on the aromatic radical cation, is thought to occur as shown. Another reaction occurring by way of initial electron transfer is the photocyclization of *N*-allylpyridinium salts; this provides a ready route to the indolizidine skeleton (3.79) in which the aromaticity of the ring is lost completely.

Most of the photochemical reactions of heteroaromatic *N*-oxides can be rationalized on the basis of an initial internal cyclization to give an oxaziridine, a three-membered heterocycle containing both oxygen and nitrogen (3.80). Ring-opening of such an intermediate would lead to a 1,2-oxazepine, containing a seven-membered ring

NH hν NH +I• NH I• NH 57%

(3.77)

OMe Cl N H O MeO hν OMe Cl •+ N H O⁻ MeO

OMe +Cl⁻ •+ N H O MeO −H⁺ OMe O NH MeO 70%

(3.78)

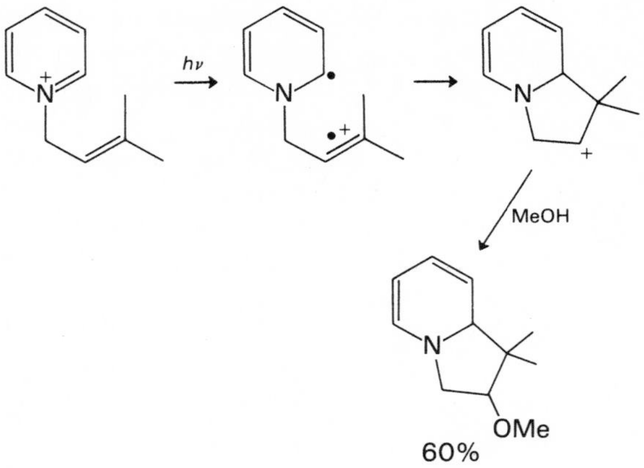

(3.79)

60%

with adjacent oxygen and nitrogen atoms, but this is observed only rarely, for example with certain acridine *N*-oxides (3.81). More usual is the production of 1,3-oxazepines, as with phenyl-substituted pyridine *N*-oxides (3.82), and these are thought to be formed by isomerization of the bicyclic oxaziridine to an epoxide of the parent pyridine.

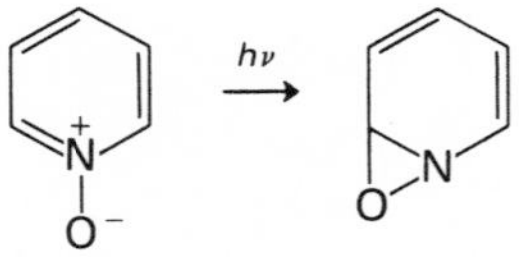

(3.80)

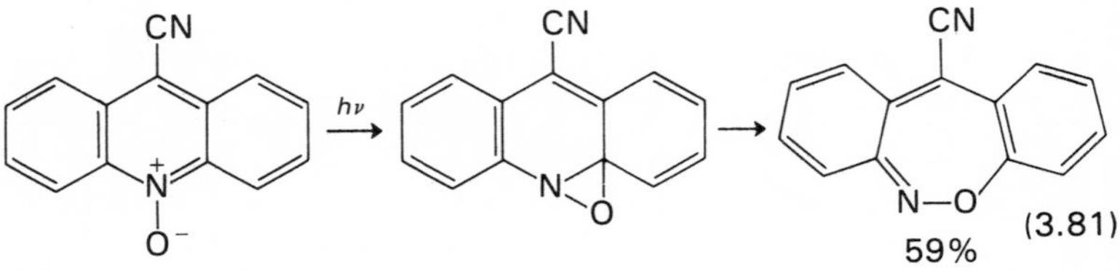

59% (3.81)

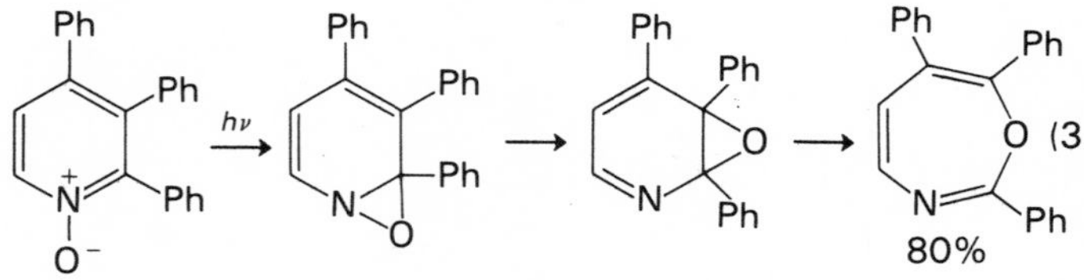

80% (3.82)

In hydroxylic solvents an alternative pathway is followed that involves ring-opening accompanied by a proton shift to yield a

2-pyridone, or, from the related quinoline *N*-oxide, 2-quinolone (3.83).

$$\text{quinoline } N\text{-oxide} \xrightarrow[\text{MeOH}]{h\nu} \text{2-quinolone (70\%)} \tag{3.83}$$

Further reading

R. A. Rossi and R. H. de Rossi, *Aromatic Substitution by the $S_{RN}1$ Mechanism*, American Chemical Society (1983). This extensive review includes a discussion of photochemical nucleophilic substitutions and their mechanisms.

R. S. Davidson, J. W. Goodwin and G. Kemp, *Advances in Physical Organic Chemistry*, vol. 20, Academic Press (1984), p. 191. The photochemistry of aryl halides and related compounds is reviewed, including substitution and cyclization reactions.

D. Bryce-Smith and A. Gilbert, in P. de Mayo (ed.), *Rearrangements in Ground and Excited States*, vol. 3, Academic Press (1980). Photochemical rearrangements of the benzene ring are described, and theoretical aspects of the mechanism are discussed.

A. Padwa, in P. de Mayo (ed.), *Rearrangements in Ground and Excited States*, vol. 3, Academic Press (1980). This review looks at the photochemical rearrangements of heterocyclic compounds with a five-membered ring.

A. Gilbert, in W. M. Horspool (ed.), *Synthetic Organic Photochemistry*, Plenum (1984). Photoaddition, photocycloaddition and photocyclization processes of aromatic compounds are all covered in this account of synthetic aspects of aromatic photochemistry.

F. B. Mallory and C. W. Mallory, *Organic Reactions*, vol. 30, Wiley (1984). This is an extensive compilation and discussion of the photocyclization reactions of stilbene and related systems.

CHAPTER 4

Photochemistry of organic carbonyl compounds

Ketones feature prominently in accounts of the earliest systematic studies of both synthetic and mechanistic organic photochemistry. They absorb in the more readily accessible wavelength ranges of the ultraviolet, so that some aromatic ketones react even in sunlight, and many of their photochemical reactions lead efficiently to products that can be easily isolated or analysed. Ketones continue to be used as substrates for many studies in photochemistry, and in general they have been more intensively studied than any other group of organic compounds.

Aliphatic ketones absorb weakly ($\epsilon \sim 20\ l\ mol^{-1}\ cm^{-1}$) at around 280 nm as a result of an $n \rightarrow \pi^*$ transition that is forbidden on both symmetry and overlap grounds. Intersystem crossing from the (n,π^*) singlet state to the corresponding triplet state is generally efficient, and because the energy and electron distribution of the (n,π^*) triplet state are not very different from those of the corresponding singlet, the two states often react in the same overall manner, though with different rate constants. The reactions observed on direct irradiation of saturated ketones may be singlet-derived, triplet-derived, or a mixture of both. For conjugated, unsaturated ketones (α,β-enones or aryl ketones), the (n,π^*) and (π,π^*) singlet states are closer in energy, and the lowest energy excited state, especially in the triplet manifold, may be (π,π^*). This is important, because photochemical reaction normally occurs through the lowest state, and the radical-like properties of (n,π^*) states are not shared by (π,π^*) states. This means that the type of product or the efficiency of reaction can be affected quite drastically by substituents that influence the relative energy levels, and even by the nature of the solvent for borderline cases.

Intersystem crossing is very efficient for aryl ketones, so that most of their photochemical reactions are the result of triplet-state processes. This efficient formation of triplet states, together with the long-wavelength absorption and small singlet–triplet energy difference, makes such an aryl ketone a good candidate for use as a triplet sensitizer, to generate the triplet states of other compounds that do not give their own triplets readily by intersystem crossing after direct absorption.

Aldehydes are quite similar to ketones in their excited state properties and photochemistry, but carboxylic acids and most of their derivatives are different. Their $n \rightarrow \pi^*$ absorption is at much shorter wavelength, and, although the lowest singlet states are (n,π^*) in nature, the lowest triplet states are often (π,π^*). The carboxylic acid compounds turn out to be generally less reactive photochemically with respect to processes that affect the C=O group, although interesting exceptions are found amongst dicarboximides (containing the grouping –CO–NR–CO–), whose photochemistry contains a range of reactions much more typical of ketones than of amides.

The major classes of reaction for the excited states of organic carbonyl compounds are homolytic cleavage of a neighbouring single bond (usually the alpha bond, (O=)C–C), abstraction of a hydrogen atom from an added compound (leading to overall photoreduction or photoaddition) or from within the molecule (leading to overall cleavage or cyclization), addition to a multiple carbon–carbon bond (most commonly resulting in cycloaddition), or rearrangement in cyclic enone or dienone systems. The final section of the chapter deals with organic thiocarbonyl compounds, the sulfur analogues of ketones and so on, which exhibit some photochemical characteristics different from those of their oxygen counterparts.

Bond cleavage

The chemical properties of an (n,π^*) excited state of a ketone can be rationalized qualitatively on the basis of a simple model showing an unpaired electron in a non-bonding orbital on oxygen and an unpaired electron in a π^* antibonding orbital. Using this model, the ready cleavage of (n,π^*) states at the alpha bond to generate an acyl radical and an alkyl radical (4.1) parallels the cleavage of an alkoxy radical, which also has an unpaired electron in a non-bonding orbital on oxygen and breaks down to give a ketone and an alkyl radical (4.2). This photochemical cleavage of ketones is often referred to as a

Norrish type 1 reaction, named after one of the pioneers of mechanistic investigations in organic photochemistry.

$$R_2C{=}\dot{\ddot{O}}\!: \longrightarrow R{-}\dot{C}{=}O + \dot{R} \qquad (4.1)$$

$$R_3C{-}\ddot{\dot{O}}\!: \longrightarrow R_2C{=}O + \dot{R} \qquad (4.2)$$

For an acyclic saturated ketone the outcome of alpha cleavage is generally loss of carbon monoxide (decarbonylation), since the acyl radical fragments to CO and a second alkyl radical. The alkyl radicals may combine or disproportionate, so that the hydrocarbon product from photolysis of acetone in the gas phase is ethane (4.3), whereas pentan-3-one gives rise to ethane, ethylene and butane (4.4). There is a fine balance of energy in the cleavage step for these ketones, and at lower temperatures (less vibrational energy) or in liquid solution (more rapid vibrational deactivation) the efficiency of decarbonylation is markedly reduced, so much so that acetone is a good solvent for many photochemical reactions at room temperature.

$$CH_3COCH_3 \xrightarrow{h\nu} CO + 2\dot{C}H_3 \longrightarrow C_2H_6 \qquad (4.3)$$

$$CH_3CH_2COCH_2CH_3 \longrightarrow CO + 2CH_3\dot{C}H_2 \longrightarrow CH_2{=}CH_2 + CH_3CH_3 \text{ or } CH_3CH_2CH_2CH_3 \qquad (4.4)$$

For related reasons, and because their excited-state energies are lower than for dialkyl ketones, diaryl ketones and simple alkyl aryl ketones do not fragment on irradiation in solution, even at higher temperatures. This leads to a photostability that is one factor contributing to the successful employment of ketones such as benzophenone (Ph_2CO) or acetophenone (PhCOMe) as triplet sensitizers. α-Cleavage for ketones in solution at room temperature is promoted if structural factors cause the bond adjacent to the carbonyl group to be somewhat weaker than normal. Hence *t*-alkyl ketones give decarbonylation products readily (4.5), as do benzyl ketones (4.6) and benzoin derivatives (4.7).

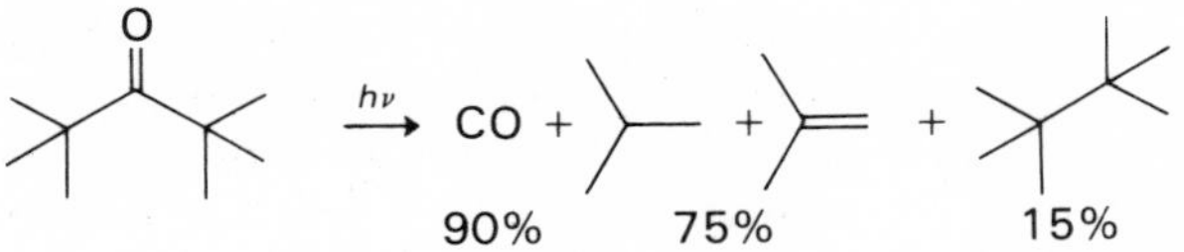

O $\xrightarrow{h\nu}$ CO + + + (4.5)

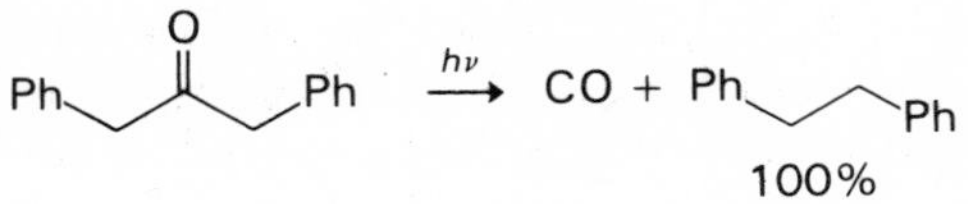

90% 75% 15%

Ph O Ph $\xrightarrow{h\nu}$ CO + Ph Ph (4.6)

100%

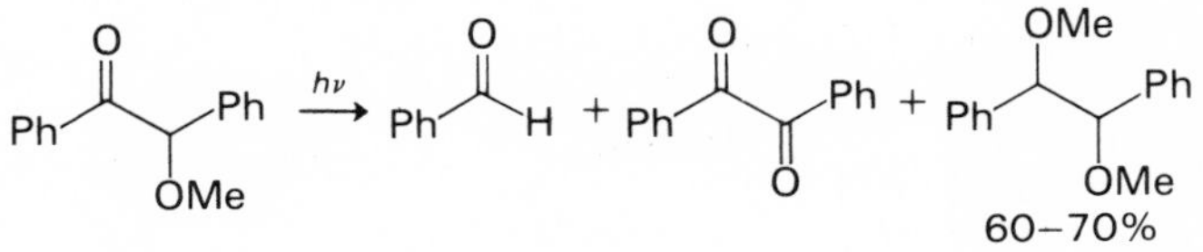

O Ph Ph OMe $\xrightarrow{h\nu}$ O Ph H + O Ph Ph O + OMe Ph Ph OMe (4.7)

60–70%

Benzoin derivatives are used as initiators for the photochemical curing of printing inks, lacquers and other surface coatings, since the intermediate radicals in a reaction such as (4.7) can be diverted to initiate the polymerization of vinyl monomers. The use of an unsymmetrical ketone (4.8) also shows that discrete radicals are produced in the cleavage reaction, since the ratio of hydrocarbon products is close to that expected for a random combination of separated radicals. Esters that give rise to similar stabilized radicals undergo loss of carbon dioxide (decarboxylation) by a closely related mechanism, and this has proved useful in making quite strained cyclic systems by irradiation of readily prepared cyclic diesters (4.9).

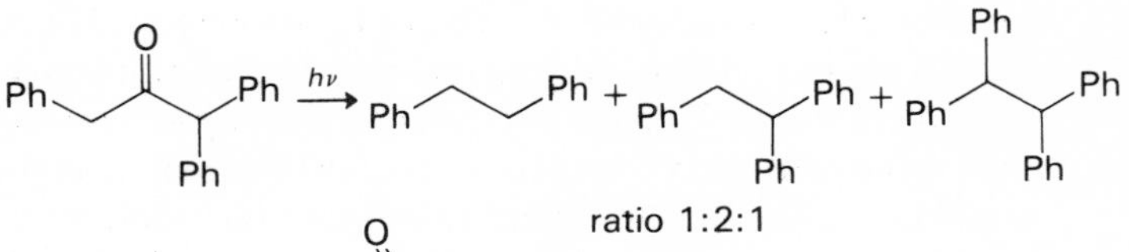

Ph O Ph Ph $\xrightarrow{h\nu}$ Ph Ph + Ph Ph Ph + Ph Ph Ph Ph (4.8)

ratio 1:2:1

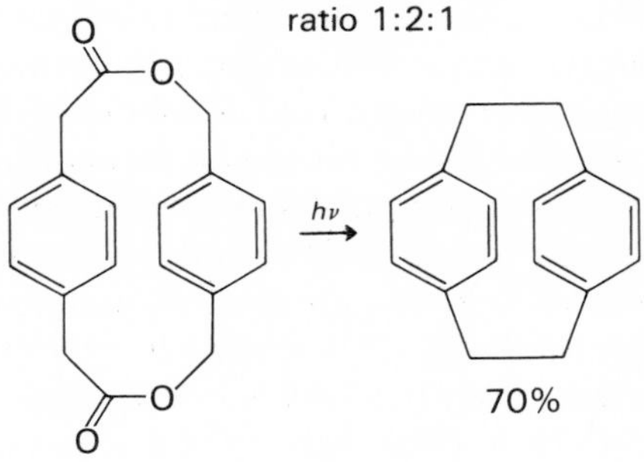

O O O O $\xrightarrow{h\nu}$ (4.9)

70%

Many carboxylic acids lose carbon dioxide on either direct or sensitized irradiation, and in some cases (4.10) the evidence points to the operation of an initial electron-transfer mechanism rather than primary α-cleavage. Cleavage occurs readily with acyl halides, and this can lead to overall decarbonylation (4.11). Aldehydes also cleave readily, since the (O=)C−H bond is more prone to homolysis than the (O=)C−C bond. This offers a convenient method for replacing the aldehydic hydrogen by deuterium in aromatic aldehydes (4.12), and a similar initial reaction step accounts for the production of chain-lengthened amides when formamide is irradiated in the presence of a terminal alkene (4.13).

$$PhOCH_2COOH \xrightarrow[Ph_2CO]{h\nu} PhOMe + CO_2 \quad (4.10)$$

$$C_3F_7C(=O)Br \xrightarrow{h\nu} CO + C_3F_7Br \ (94\%) \quad (4.11)$$

$$PhCHO \xrightarrow[D_2O,\ Me_2CO]{h\nu} PhCDO \quad (4.12)$$

$$HCONH_2 + CH_3(CH_2)_5CH{=}CH_2 \xrightarrow{h\nu} CH_3(CH_2)_7CONH_2 \ (45\%) \quad (4.13)$$

β,γ-Unsaturation promotes cleavage of the alpha bond because the alkyl radical is stabilized by allylic resonance. For many β,γ-enones recombination of the radicals can compete with loss of carbon monoxide from the acyl radical, so that the overall reaction is one of isomerization, as seen for a cyclic system in (4.14). Triplet sensitization of β,γ-unsaturated ketones may lead to a different product incorporating a cyclopropyl ketone unit (4.15); this is obtained by way of a 1,2-shift of the acyl group and is the oxygen counterpart of the di-π-methane reaction (see p. 54), sometimes called the oxa-di-π-methane rearrangement. 1,3-Shifts also occur in vinyl esters to give β-diketones (4.16) and in phenyl esters to give phenols with acyl substituents in the ring (see p. 84).

With saturated, or non-conjugated unsaturated, cyclic ketones, α-cleavage takes place readily in solution on irradiation, usually in a triplet-state process. The resulting acyl–alkyl biradicals have a num-

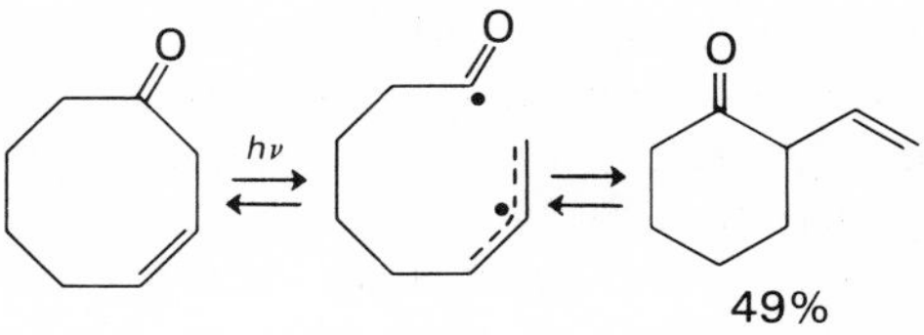

49%

(4.14)

>60%

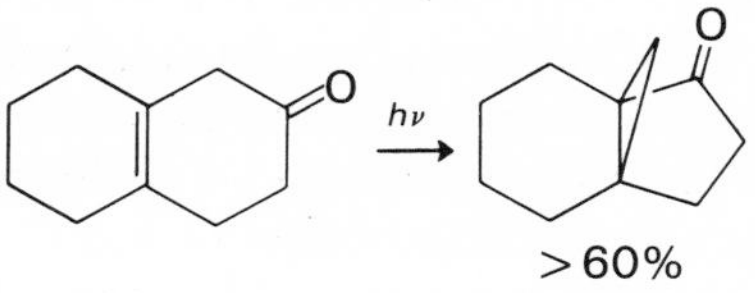

(4.15)

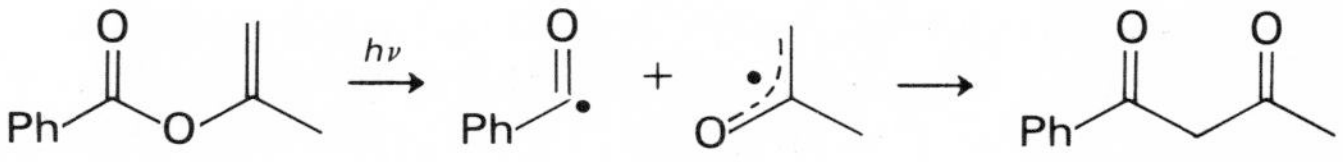

(4.16)

ber of pathways open to them. Loss of carbon monoxide leading to overall decarbonylation may occur, though under normal conditions this is efficient only if stabilized alkyl radicals are involved (4.17). More commonly the carbonyl group is retained in one of four processes. First, the biradical may recombine to give the original substrate; this can be an energy-wasting process, but it is chemically useful if the α-carbon is chiral, because an epimer of the substrate

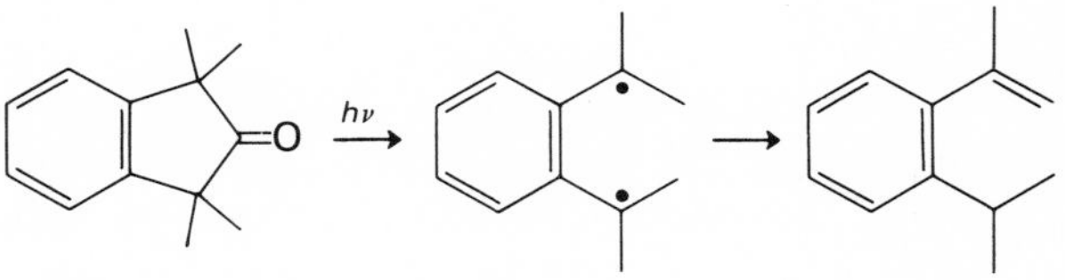

(4.17)

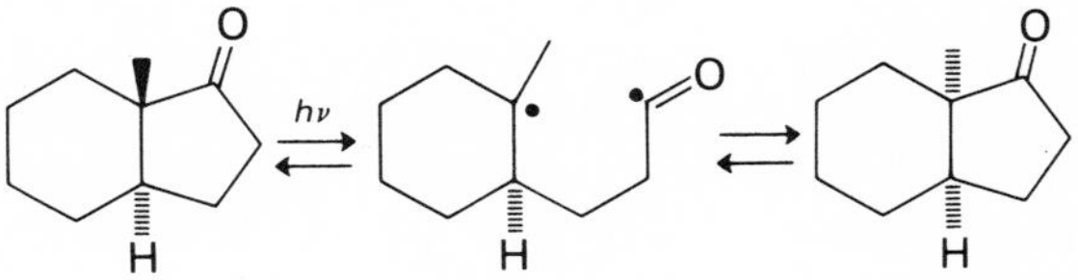

(4.18)

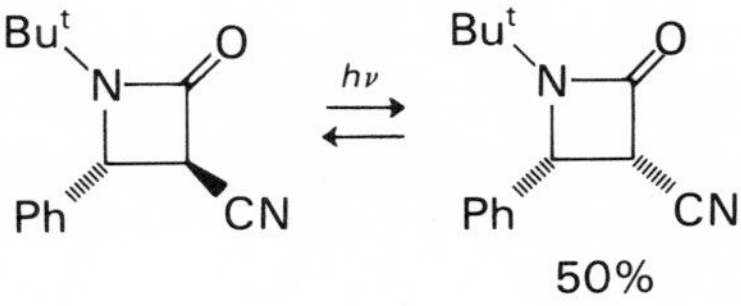

50%

(4.19)

may also be formed. This provides a method for isomerization of ketones with a quaternary α-carbon (4.18), or sensitive compounds such as β-lactams (4.19), neither of which are amenable to epimerization by the more commonly used method with a base. The reaction is useful generally for enriching an epimeric mixture in favour of the less stable isomer. When the α-position carries a cyclopropyl substituent, recombination after rearrangement of the alkyl radical offers a neat way of ring-expansion for cyclic ketones by three carbon atoms, and the process has been used (4.20) in a synthesis of a fifteen-membered cyclic ketone used in fragrances.

The acyl–alkyl biradical obtained by ring-opening of a cyclic ketone is able to undergo intramolecular disproportionation in one of two ways. A hydrogen atom may be transferred to the acyl radical from the position adjacent to the alkyl group, and this produces an unsaturated aldehyde (4.21). Alternatively, a hydrogen may be transferred to the alkyl radical from the position adjacent to the acyl group, and this results in the formation of a ketene (4.22). Many ketenes are labile, and the use of a nucleophilic solvent or addend,

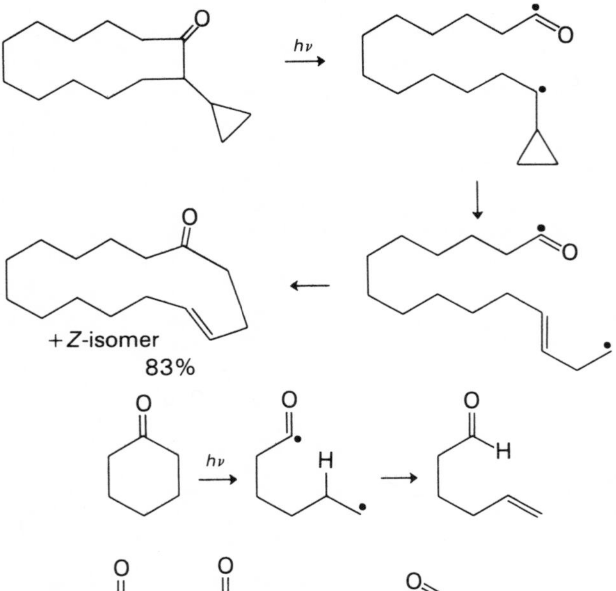

$h\nu$

+ Z-isomer
83%

(4.20)

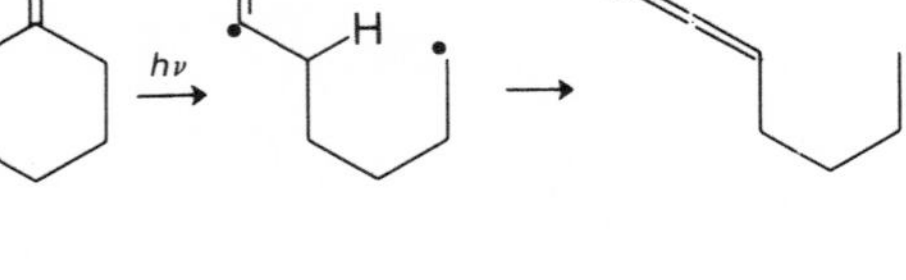

$h\nu$

(4.21)

$h\nu$

(4.22)

such as water, alcohol or amine, ensures that the ketene is trapped efficiently, as a carboxylic acid, ester or amide, respectively. The competition between the two modes of disproportionation for the biradical depends on such factors as ring size and position of substituents, and it can be rationalized by consideration of the conformations required in the intermediate biradical. Five-membered rings give largely unsaturated aldehyde (or related) products, as exemplified by the reaction of 2-(ethoxycarbonyl)cyclopentanone (4.23), the β-ketoester obtained by Dieckmann cyclization of diethyl hexanedioate. Similar reactions have been employed in the synthesis of acyclic compounds with remote unsaturation (4.24), many of which are related to naturally occurring terpenes, insect pheromones, or fragrances, and also in the formation (4.25) of a key intermediate in prostaglandin synthesis. Notice that for all the systems shown, α-cleavage in an unsymmetrical ketone occurs on the more heavily substituted side of the carbonyl group, as would be expected in the light of relative biradical stabilities.

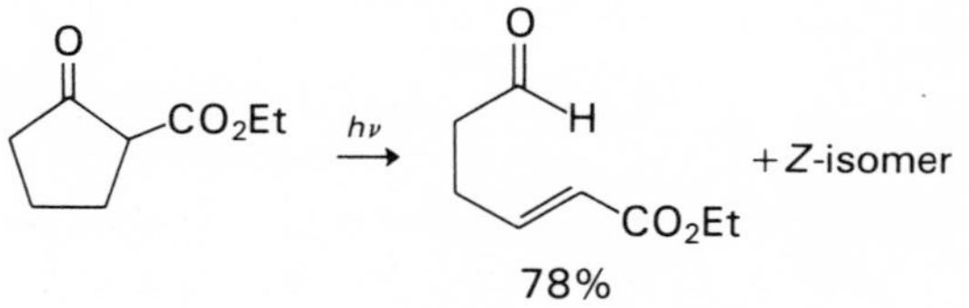

CO_2Et $h\nu$ → H CO_2Et 78% + *Z*-isomer (4.23)

$h\nu$ → H + *Z*-isomer 80% (4.24)

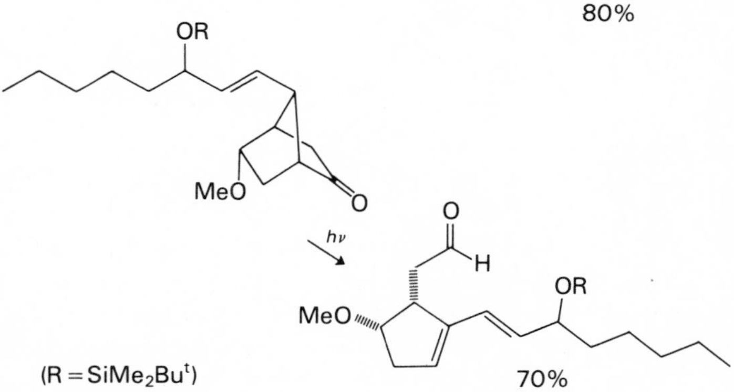

OR MeO O $h\nu$ → O H OR MeO 70% (4.25)

(R = $SiMe_2Bu^t$)

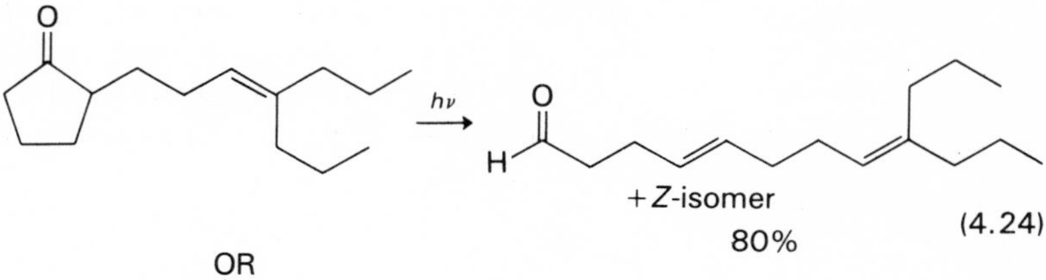

Ketene formation is more important for six-membered cyclic ketones, and irradiation of a 2-alkylcyclohexanone in water gives a straight-chain carboxylic acid (4.26). If ketene formation is prevented by structural factors, an unsaturated aldehyde may be formed in good yield (4.27). Larger-ring ketones do sometimes give products by α-cleavage followed by internal disproportionation, but intramolecular hydrogen abstraction if often a successful competing process (see p. 122).

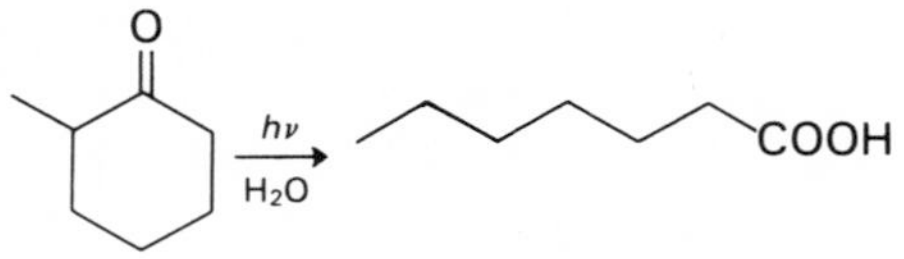

$h\nu$, H_2O → COOH (4.26)

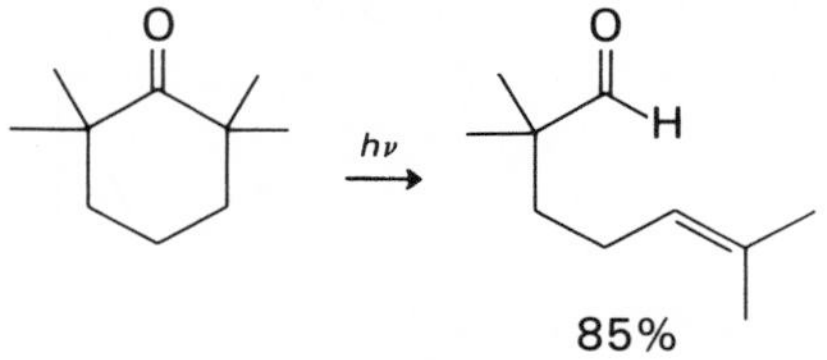

$h\nu$ → H 85% (4.27)

Smaller-ring ketones, especially cyclobutanones and more rigid cyclopentanones or cyclohexanones, give biradicals that follow the fourth of the pathways in which carbon monoxide is not lost. In this process a new oxygen–carbon bond is formed by attack of the oxygen of the acyl radical on the alkyl radical centre; this generates a carbene which can subsequently react with a nucleophilic solvent such as methanol (4.28).

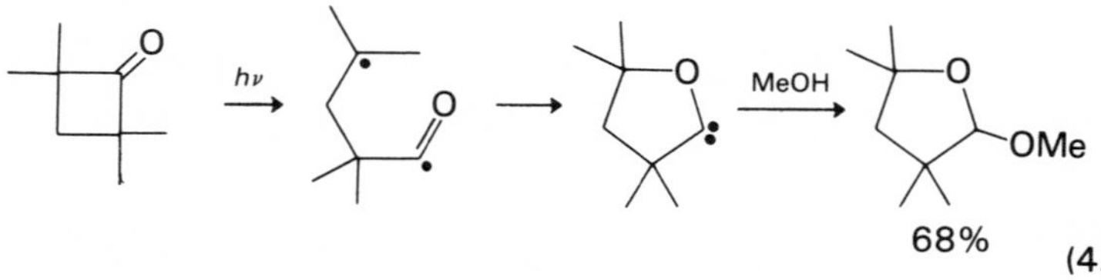

$h\nu$ → → MeOH → OMe 68% (4.28)

The examples of photochemically induced bond-cleavage that have been described so far in this section have been α-cleavage processes, but cleavage of the C(α)–C(β) bond may occur if this bond is especially weak because of substituents or structural features. Hence 2-chloro carboxylate esters can be converted to the parent esters by irradiation in a hydrogen donor solvent (4.29), and *N*-chlorosuccinimide undergoes photoaddition with an alkene (4.30) by initial cleavage of the nitrogen–chlorine β-bond. This type of cleavage is very efficient for diacyl peroxides (RCO.OO.COR), which makes them a very useful source of radicals ($RCO_2^{\cdot}$ or R·) for the initiation of radical chain processes.

Cl COOEt $\xrightarrow[c\text{-}C_6H_{12}]{h\nu}$ COOEt >90% (4.29)

O N—Cl O + $\xrightarrow{h\nu}$ Cl N O O 54% (4.30)

Hydrogen abstraction

Just as (n,π^*) excited states of ketones exhibit radical-like behaviour in their α-cleavage reactions, so they resemble alkoxy radicals in their ability to abstract a hydrogen atom from a suitable donor (4.31). The initially formed pair of radicals may combine to give an overall photoaddition product, or the radical derived from the ketone may dimerize, or abstract a second hydrogen atom, resulting in overall photoreduction of the carbonyl compound.

$$>C=O^* + RH \longrightarrow R^{\bullet} + >\dot{C}-OH \qquad (4.31)$$

R C OH; OH C—C OH; H C OH

The hydrogen donor (RH) may be an alkane, but often the use of a saturated hydrocarbon leads to inefficient reaction; the balance of energy is quite fine, and for the reaction to be efficient a compound with a slightly weaker C–H bond is required. Suitable donors are primary or secondary alcohols or ethers (C–H adjacent to oxygen), alkylbenenes (benzylic C–H), or certain alkenes (with allylic C–H); these all give radicals that are stabilized relative to the corresponding alkyl radicals. One outcome of the facile reaction between excited-state ketones and these compounds is that some of the commonly used solvents for thermal reactions, such as ethanol, diethyl ether or tetrahydrofuran, are often not appropriate for use for the photochemical reactions of carbonyl compounds.

The overall course of reaction depends on the relative rate constants for the various secondary radical processes. Aliphatic ketones are often photoreduced to secondary alcohols (4.32), but although there are interesting features in the stereochemistry of the reduction, the method is not a worthwhile alternative to thermal reduction using hydride reagents, except in cases where the substrate is sensitive to basic conditions. Photoaddition of methanol is promoted in the presence of titanium(IV) chloride, both for acyclic and cyclic (4.33) ketones; the titanium involvement probably starts in the early steps of the reaction, but the detailed mechanism is not known. Addition may also be a major pathway when cyclohexene is used as hydrogen source (4.34); unlike many other simple alkenes, cyclohexene does not readily give oxetanes by photocycloaddition (see p. 126).

Aryl ketones such as benzophenone (Ph_2CO) or acetophenone

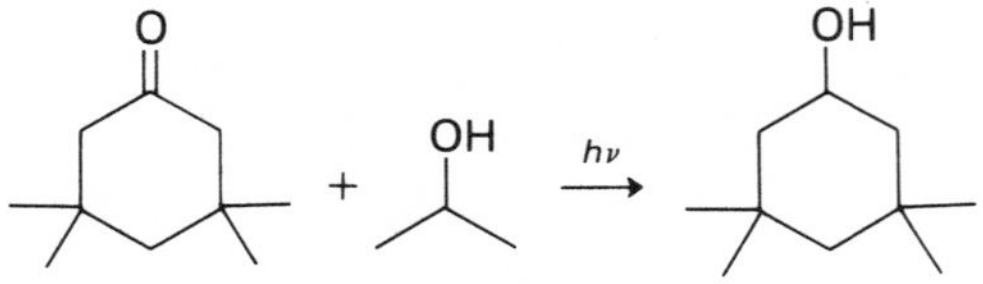

O + OH $\xrightarrow{h\nu}$ OH (4.32)

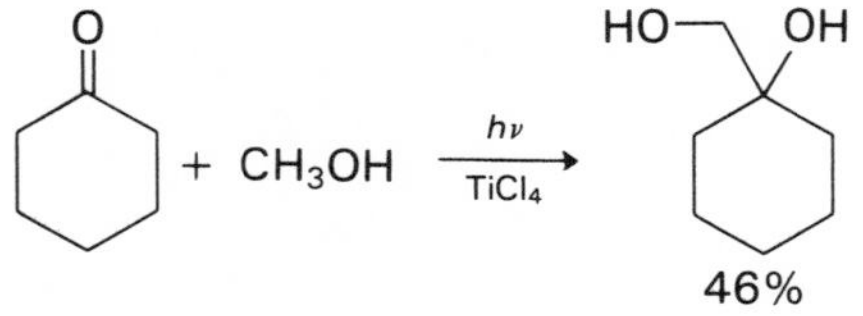

O + CH_3OH $\xrightarrow[TiCl_4]{h\nu}$ HO OH (4.33)

46%

O + $\xrightarrow{h\nu}$ OH (4.34)

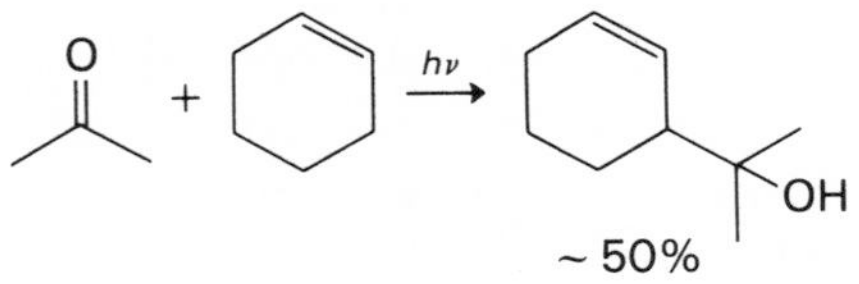

~ 50%

(PhCOMe) give good yields of the dimeric reduction product, a pinacol, on irradiation in methanol or propan-2-ol (4.35), and this is a convenient method for making such symmetrical pinacols. The mechanism has been studied in detail, and only one photon is required to generate two aryl ketyl radicals, since the radical derived from the aliphatic alcohol exchanges a hydrogen atom with a second molecule of ground-state aryl ketone (4.36). Support for this pathway is seen in the predominance of a photoaddition product (4.37) when an extremely dilute solution of benzophenone in methanol is irradiated—under these conditions the rate of reaction between $\dot{C}H_2OH$ and ground-state benzophenone is markedly reduced. Pinacols are less readily formed at higher temperatures, and the photochemical reaction of benzophenone in propan-2-ol under reflux leads instead to reduction to the secondary alcohol, Ph_2CHOH.

$$Ph_2C{=}O + (CH_3)_2CHOH \xrightarrow{h\nu} HO{-}CPh_2{-}CPh_2{-}OH \quad (100\%) \tag{4.35}$$

$$(CH_3)_2\dot{C}OH + Ph_2C{=}O \xrightarrow{h\nu} (CH_3)_2C{=}O + Ph_2\dot{C}OH \tag{4.36}$$

$$Ph_2C{=}O\ (10^{-4}\ \text{M}) + CH_3OH \xrightarrow{h\nu} HOCH_2{-}CPh_2{-}OH \quad (>90\%) \tag{4.37}$$

For the reactions described so far in this section, the ketone substrates have lowest excited states that are (n,π^*) in character; aliphatic ketones may react by way of the singlet or the triplet state, and aryl ketones normally through the triplet because intersystem crossing is very efficient. The efficiency of photochemical hydrogen abstraction from compounds such as alcohols or ethers is very much lower if the ketone has a lowest (π,π^*) triplet state, as does 1- or 2-acetylnaphthalene ($C_{10}H_7COMe$). However, all aryl ketones, regardless of whether their lowest triplet state is (n,π^*) or (π,π^*), react photochemically with amines to give photoreduction or photoaddition products. A different mechanism operates (4.38), that begins

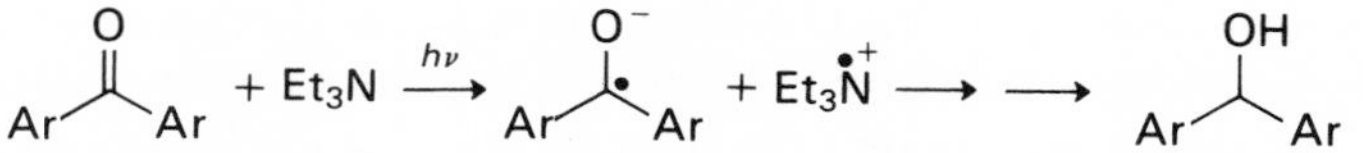

$$\mathrm{Ar_2C{=}O} + \mathrm{Et_3N} \xrightarrow{h\nu} \mathrm{Ar_2\dot{C}{-}O^-} + \mathrm{Et_3\overset{\bullet +}{N}} \longrightarrow \longrightarrow \mathrm{Ar_2CH{-}OH} \quad (4.38)$$

with electron transfer from the amine to the excited ketone giving radical ions (in some instances the radical ions have been detected spectroscopically); later steps include proton transfer to the radical anion. All excited states have enhanced electron-acceptor properties, regardless of their electron distribution, and this mechanism provides an alternative route to products that might in principle arise by way of hydrogen-atom abstraction.

The photoreduction of carboxylic acids and their derivatives has not been widely reported, although cyclic imides do behave like ketones in such reactions (4.39). Quinones are readily photoreduced, and this is of particular importance for anthraquinone dyes or pigments used on cellulose materials such as cotton. Some quinone dyes cause what is known as phototendering—the fabric progressively weakens and can be torn easily after exposure of the dyed

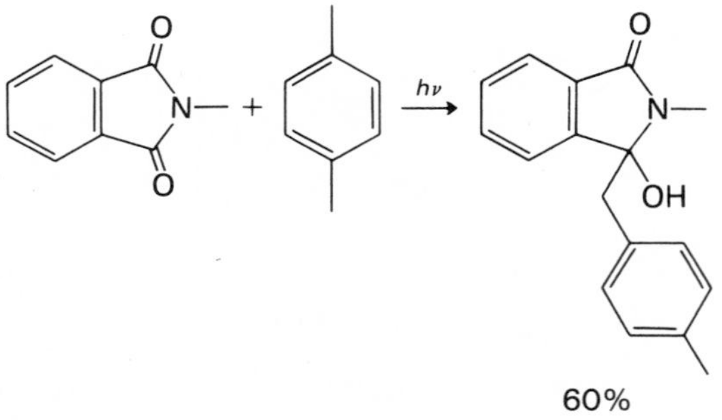

O N— O + $\xrightarrow{h\nu}$ O N— OH

60%

(4.39)

HO R O O HO O R OH H $\xrightarrow{h\nu}$ HO R O O HO O R OH $\longrightarrow$ HO R O O HO OH O + R$^\bullet$

+ O O OH O

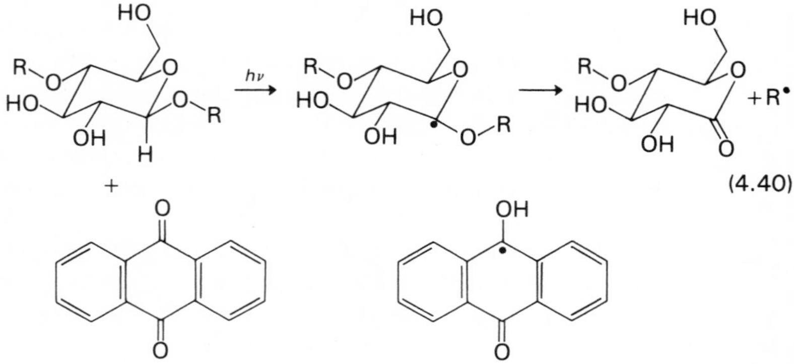

(4.40)

material to light. The chemical reaction causing this effect starts with photochemical hydrogen abstraction by excited-state quinone from a carbohydrate unit of the material (4.40); radical cleavage causes a break in the polymer chain, and the macroscopic manifestation of this is loss of fibre strength. Dyes that are much less prone to cause phototendering are designed to have a readily accessible hydrogen atom within the dye molecule that can be abstracted preferentially on excitation in a reversible reaction that does not lead to overall chemical change.

Although it does not involve hydrogen abstraction, it is worth noting that some ketones add alcohol on irradiation to give an acetal (4.41). The reaction is similar to the thermal, acid-catalysed formation of acetals, except that it occurs, especially for cyclic ketones, in solutions which do not have any strong acid added. There is an element of irreproducibility in the systems, and it seems that traces of acid in the alcohol initiate the reaction, which can be prevented by adding a small amount of solid potassium carbonate to the alcohol solution.

O

MeO OMe

$h\nu$

MeOH

(4.41)

Intramolecular hydrogen abstraction

Hydrogen abstraction can occur from a position within the ketone molecule, and this generates a biradical that may cyclize by combination of the radical centres. The overall photocyclization process is observed for a wide variety of compound types, and it has been used extensively to make cyclic or polycyclic systems. In an unconstrained system a ketone (n,π^*) excited state shows a preference for abstraction from the γ-position (4.42), which can be understood on

H R′ O γ β R α

$h\nu$

O H R′ R

(4.42)

the basis of an energetic preference for reaction by way of a six-atom cyclic activated complex (transition state). γ-Abstraction generates a 1,4-biradical that can cyclize to give a cyclobutanol (4.43), but exceptionally this type of biradical (as opposed to 1,3 or 1,5 etc. biradicals) can also undergo cleavage to give an alkene and the enol of a shorter-chain ketone. The overall photoelimination reaction has been known for a long time and is often referred to as a Norrish type 2 process. It is the predominant photochemical reaction for many acyclic, dialkyl or alkyl aryl ketones that have at least three carbon atoms in one of the groups attached to the carbonyl unit (4.44), and related reactions occur for carboxylic acids or esters.

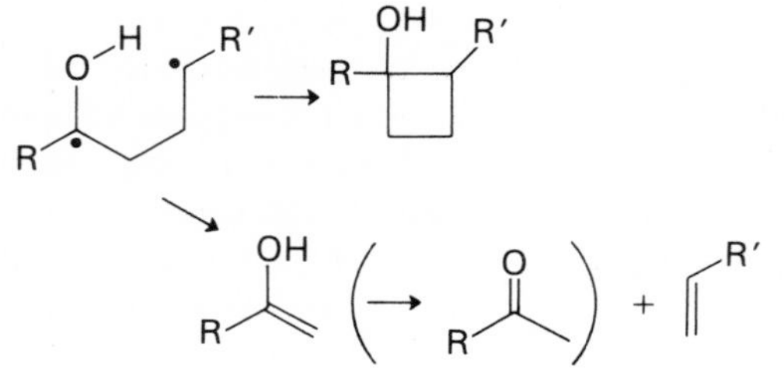
H
O
R'
R
OH
R'
R
OH
R
O
R
R'
+

(4.43)

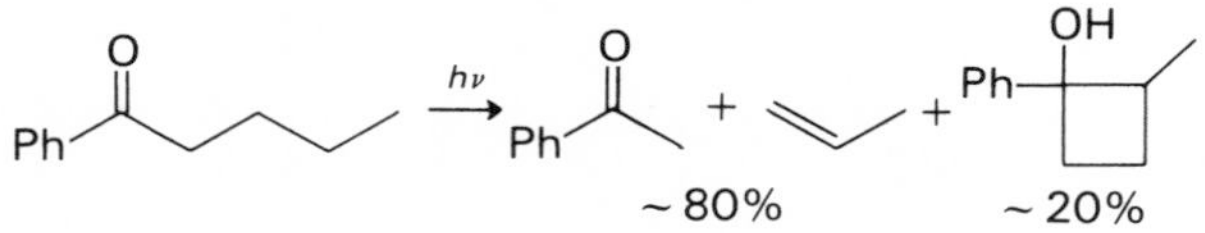
O
Ph
hν
O
Ph
+
+
OH
Ph
~80%
~20%

(4.44)

The biradical intermediates have in some cases been detected by flash spectroscopy, or trapped by added reagents such as a *t*-alkyl nitroso compound. The enol produced by cleavage of the biradical is relatively inert at low temperatures, and it can be studied spectroscopically after irradiation of the ketone in solutions cooled below −50°C. There have been many mechanistic studies of the Norrish

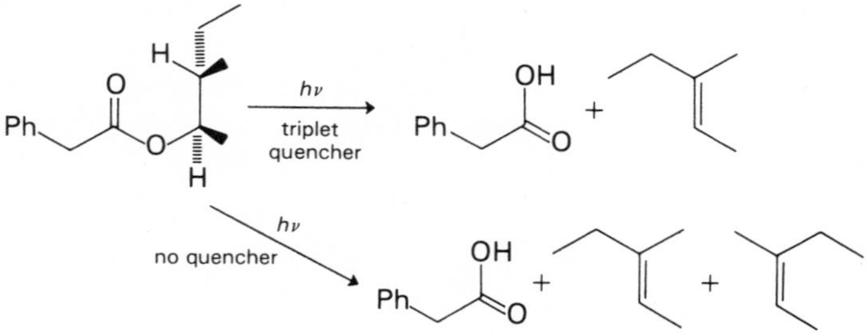
H
O
Ph
O
H
hν
triplet
quencher
OH
Ph
O
+
hν
no quencher
OH
Ph
O
+
+

(4.45)

type 2 process, and one feature worth highlighting is that the stereoselectivity depends on the multiplicity of the reactive excited state. Aryl ketones generally react by way of a triplet (n,π^*) state, but dialkyl ketones and related esters react through both the singlet and the triplet states. The singlet-derived reaction, studied in the presence of excess triplet quencher, is highly stereoselective (4.45), but the triplet-derived reaction is not.

Norrish type 2 photoelimination is not often a useful reaction in synthesis, although there are a few exceptions, such as the oxidation of a 3,4-epoxy-substituted cyclohexanol to a cyclohexanone (4.46) by way of the pyruvate ester; in this instance many conventional oxidizing agents for secondary alcohols were found to cause reaction at the epoxide unit also. Of greater interest to the synthetic chemist are the cyclization reactions, and for the unconstrained systems described previously there has been some success in increasing the proportion of cyclobutanol product (to as high as 70%) by employing more highly ordered arrays of molecules, for example in micelles or in inclusion complexes. Conformational factors undoubtedly play a large part in determining the cyclization : cleavage ratio for the biradical, largely because the two singly occupied orbitals in the biradical and the bonding orbital corresponding to the C(α)–C(β) bond that is broken on cleavage need to be aligned in a particular (parallel) way for efficient cleavage. Conformational preferences that inhibit this alignment will promote cyclization, and for this reason α-diketones (4.47) and large-ring ketones such as cyclododecanone (4.48) give high yields of cyclobutanol products.

Photocyclization to give a four-membered cyclic alcohol can be

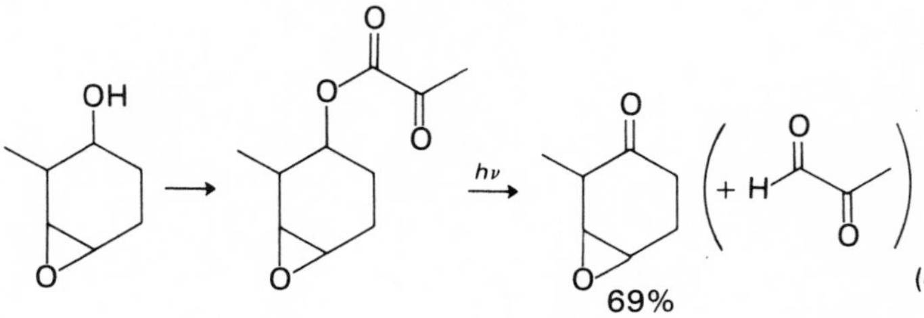

hν

69%

(4.46)

hν

100%

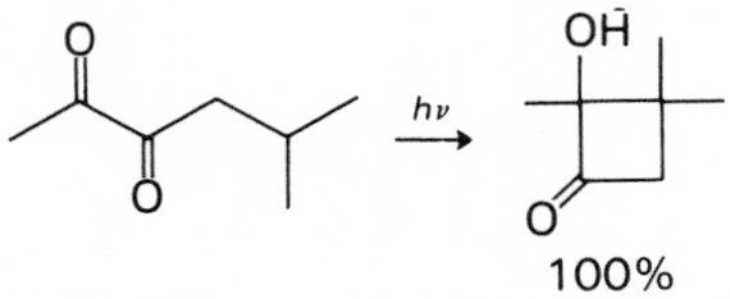

(4.47)

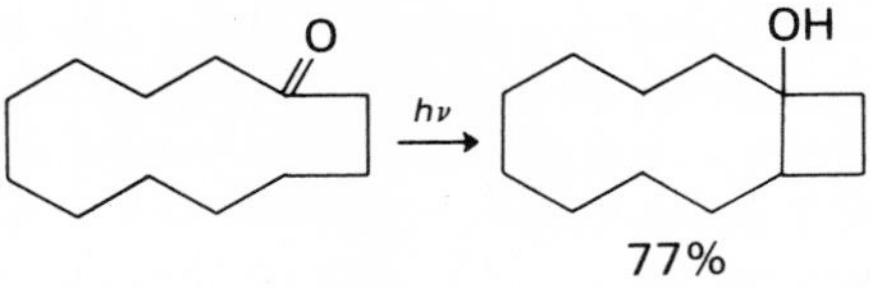

favoured by the presence of a heteroatom in the β-position, as with some *N*-substituted α-aminoacetophenones (4.49). It also occurs with reasonable efficiency in some fairly rigid systems where, apart from any conformational factors, the alkene obtained by cleavage of the biradical would be severely strained (4.50).

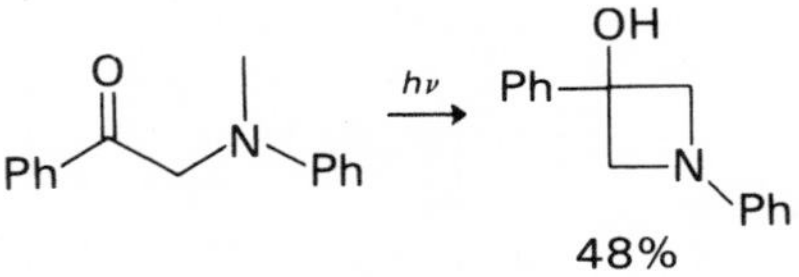

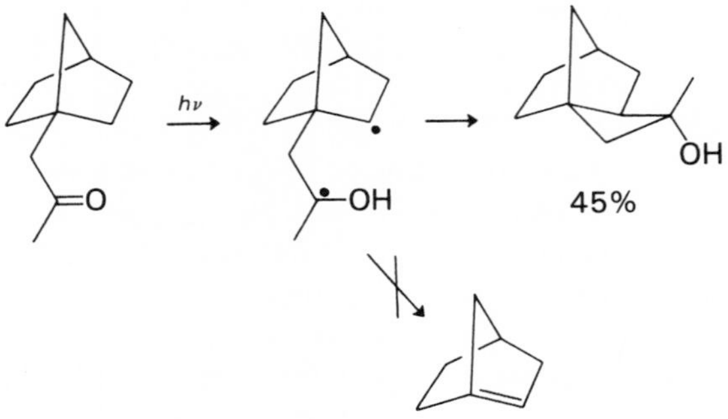

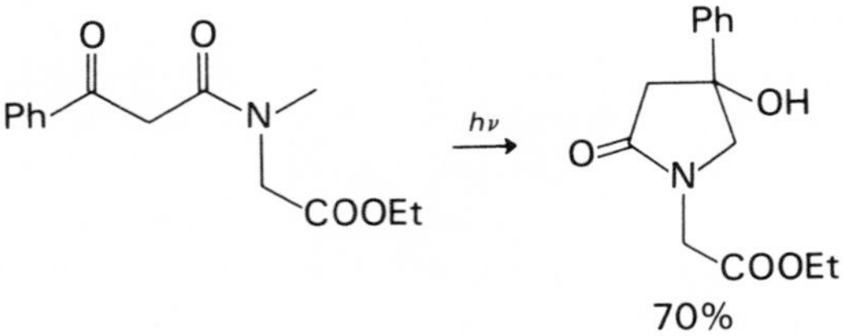

Biradicals other than 1,4-biradicals do not break down to give electron-paired cleavage products, and so photocyclization is an efficient reaction for carbonyl compouds where the γ-position does not carry a hydrogen atom. Hence *N*-methyl β-ketoamides react to give substituted pyrrolidin-2-ones (4.51), and β-allyloxyketones lead to substituted tetrahydrofurans (4.52). In both of these examples hydrogen abstraction involves the δ-position with respect to the reacting ketone group. In a few instances β-abstraction takes place to give a cyclopropanol product (4.53), even though a hydrogen is present at the δ-position. If both γ- and δ-positions are blocked by carrying no hydrogen, reaction at the ε-position may be efficient (4.54).

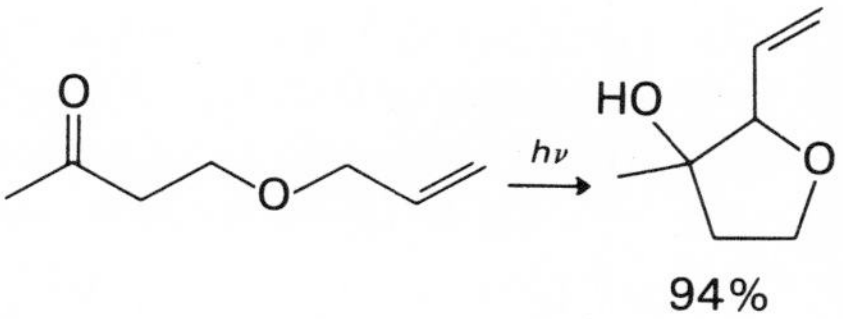

(4.52)

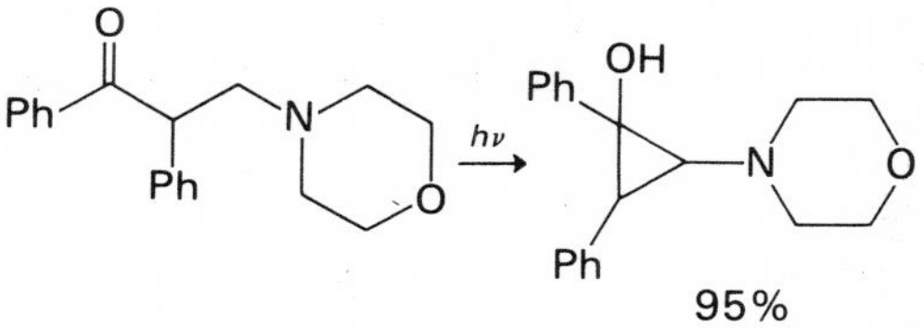

94%

(4.53)

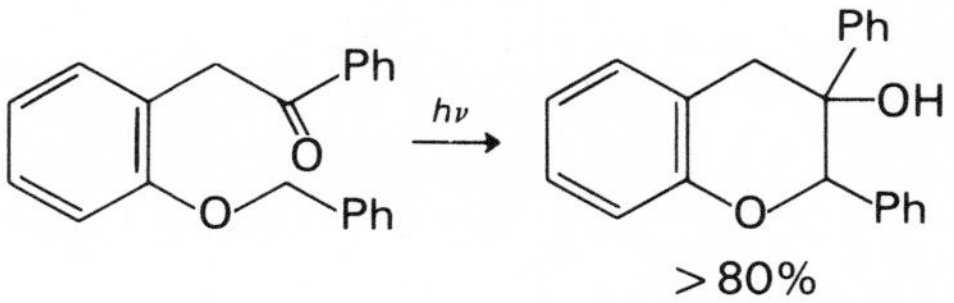

95%

(4.54)

>80%

Even in the presence of a γ-hydrogen atom available for abstraction, other factors may direct reaction to a different position. Cyclodecanone reacts to give bicyclo[4.4.0]decan-1-ol (4.55), and the main influence here is the proximity of the ε-hydrogen to the oxygen of the carbonyl group in the major conformations of the ten-membered ring. Reaction at the ε-position rather than the γ-position is evident in the reaction of some *N*-substituted phthalimides (4.56), and in this case the reason may be that electron transfer occurs as the first step in the excited state, and subsequent proton transfer and cyclization is influenced by the forces attracting the imide radical anion and amine radical cation portions of the intermediate species. Spectacular examples of this effect are seen in imides whose *N*-substituent ends in a methylthio group (sulfides are very good electron donors) and which lead to very large ring compounds in high yield (4.57).

O OH $h\nu$ 52% (4.55)

O Ph HO Ph N N Ph $h\nu$ N Ph O O 28% (4.56)

O N$(CH_2)_{12}$NHCO$(CH_2)_{10}$SMe O $h\nu$ O N O NH HO S 57% (4.57)

The lowest triplet excited state of α,β-unsaturated ketones is (π,π*) in character, and through this state *cis–trans* isomerization can occur as for alkenes (see p. 42); thus cyclo-oct-2-enone can be converted to the less stable *trans* isomer in high yield (4.58). For acyclic α,β-unsaturated ketones, acids or esters, the *cis* isomer can then undergo a second, much less efficient photochemical reaction to give the β,γ-unsaturated isomer by way of hydrogen abstraction followed by proton transfer within the enol so formed (4.59). It is thought that this reaction occurs through the singlet (*n*,π*) state. The reaction is much more efficient for both unsaturated esters and unsaturated ketones if a mild base is present (4.60), and the β,γ-unsaturated product can be formed in high yield. The action of the base is to promote isomerization of the dienol to the non-conjugated product rather than reversion to the conjugated substrate.

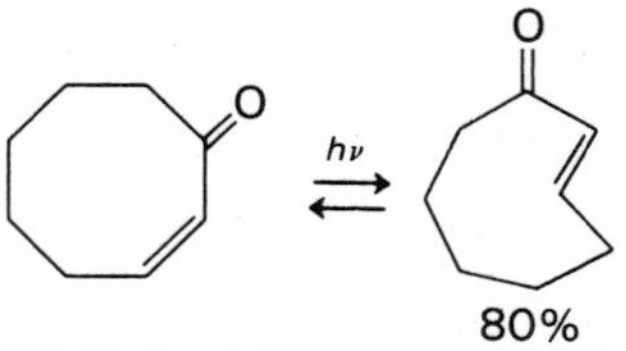

80%

(4.58)

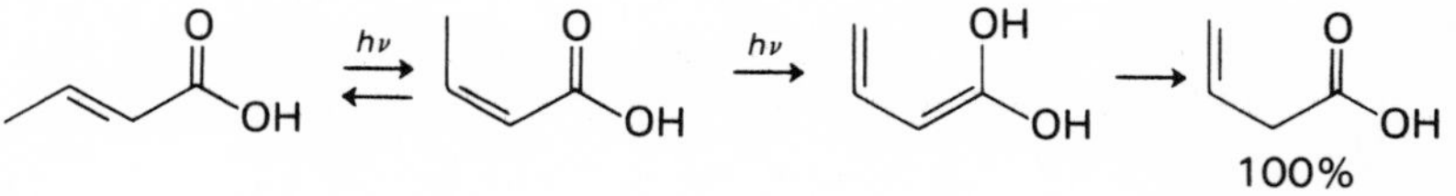

hν

hν

OH

100%

(4.59)

hν

pyridine

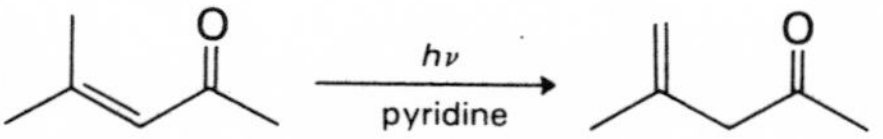

(4.60)

o-Alkylbenzophenones give similar enols on irradiation (4.61) as a result of intramolecular hydrogen abstraction, but for these compounds the enol reverts to the original ketone rather than an isomeric one. The overall result is degradation of light energy to thermal energy, and this type of reaction forms the basis for methods of protection (e.g. of plastics) against solar radiation. The photo-enol is a *cisoid* conjugated diene, and as such it can be trapped by an added dienophile such as dimethyl fumarate.

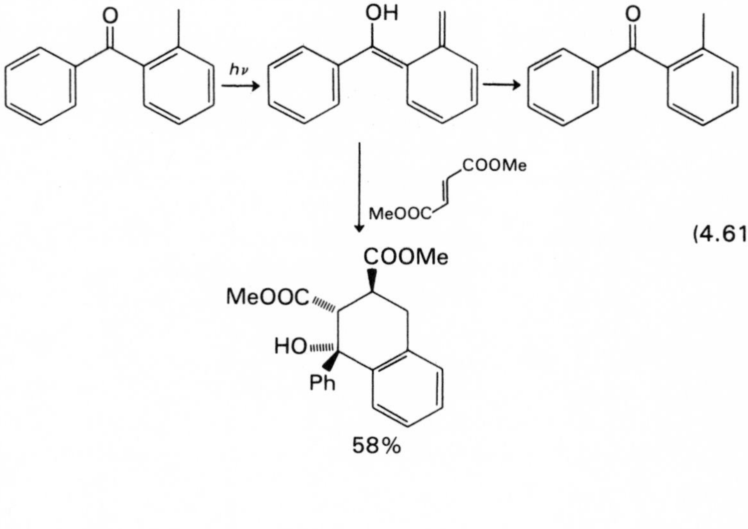

hν

COOMe

MeOOC

(4.61)

COOMe

MeOOC

HO

Ph

58%

Cycloaddition to carbon–carbon multiple bonds

The (n,π^*) excited state of a ketone has electrophilic character, similar to that associated with alkoxy radicals, and it is not surprising that these excited states readily attack carbon–carbon multiple bonds. The overall reaction that normally ensues is a cycloaddition, giving a four-membered oxygen heterocycle—an oxetane from an alkene addend (4.62), or an oxete from an alkyne addend (4.63). Some oxetanes are of interest in their own right, but many are useful intermediates in the synthesis of other compounds.

$$\mathrm{R_2C{=}O} + \mathrm{R_2C{=}CR_2} \xrightarrow{h\nu} \quad (4.62)$$

$$\mathrm{R_2C{=}O} + \mathrm{RC{\equiv}CR} \xrightarrow{h\nu} \quad (4.63)$$

When an alkyl or aryl ketone, or an aryl aldehyde, reacts with an alkyl-substituted ethylene, or with an electron-rich alkene such as a vinyl ether, the mechanism involves attack by the (n,π^*) triplet state of the ketone on ground-state alkene to generate a 1,4-biradical that subsequently cyclizes. The orientation of addition is in keeping with this proposal, since the major product is formed by way of the more stable of the possible biradicals, as seen for benzophenone and 2-methylpropene (4.64). As would be expected for a triplet-state reaction, the stereoselectivity is low, and benzophenone gives the same mixture of stereoisomers when it reacts with either *trans* or

$$\mathrm{Ph_2C{=}O} + \mathrm{(CH_3)_2C{=}CH_2} \xrightarrow{h\nu} \quad (4.64)$$

84%

9%

cis-but-2-ene (4.65), because the biradical is sufficiently long-lived for the stereochemical memory to be lost by bond rotation. Further evidence for the intermediacy of the biradical comes from the observation of a vinyl ether and an allyl ether as minor products that result from intramolecular disproportionation when acetone and 2,3-dimethylbut-2-ene are irradiated (4.66), and also the ability to form a 1,2,4-trioxane by trapping with oxygen the biradical from *p*-benzoquinone and cyclohexene (4.67).

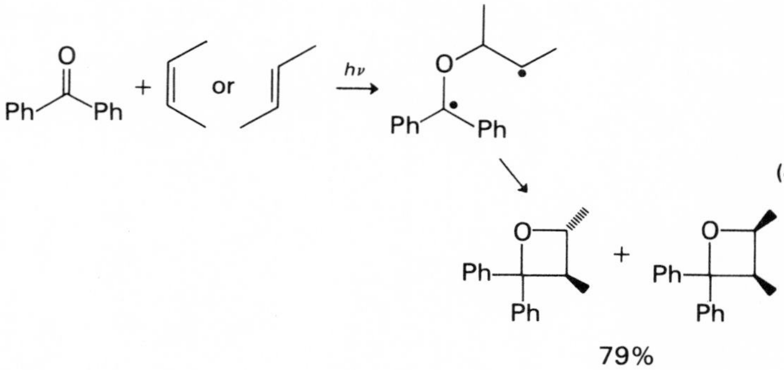

Ph Ph O + or $h\nu$ → O Ph Ph → Ph O Ph + Ph O Ph 79% (4.65)

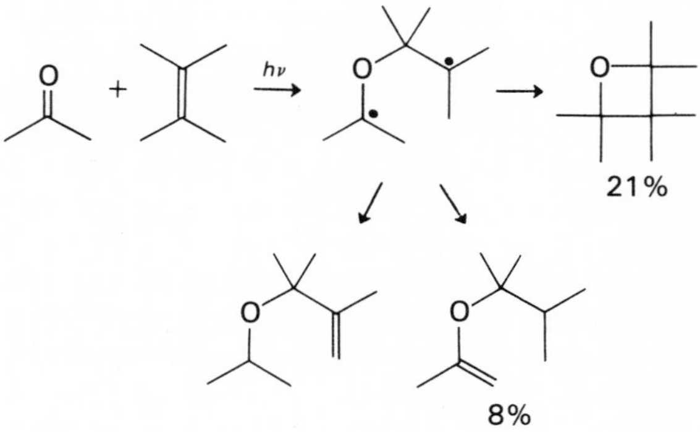

O + $h\nu$ → O → O 21% (4.66)

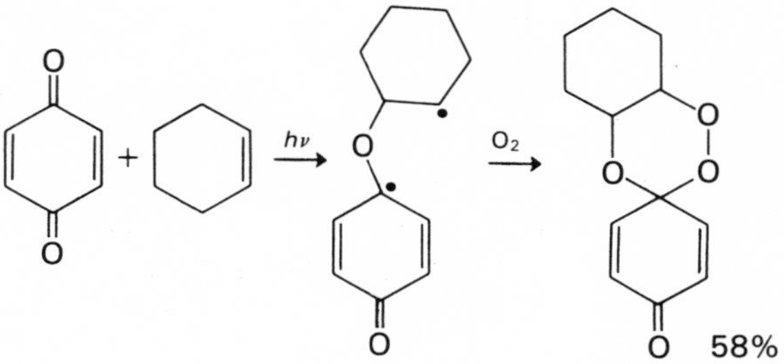

O O 8%

O O + $h\nu$ → O O O_2 → O O O O 58% (4.67)

Cyano-substituted ethylenes react in a different way with aliphatic ketones. The orientation of photochemical cycloaddition (4.68) is the opposite of that found for electron-rich alkenes, and the reaction is highly stereoselective (4.69) in the early stages. These processes involve the formation and subsequent decay of an excited complex (exciplex) from the (n,π^*) singlet state of the ketone and the alkene. Aryl ketones undergo intersystem crossing so efficiently that such a singlet-state reaction is rarely observed, but the reaction of a benzoate ester with an electron-rich alkene (4.70) may well be of this type, with the ester acting as electron-acceptor rather than electron-donor.

O + CN $\xrightarrow{h\nu}$ O CN (4.68)

42%

O + CN CN $\xrightarrow{h\nu}$ O CN CN (4.69)

90%

Ph OMe (O) + MeO OMe $\xrightarrow{h\nu}$ O OMe OMe Ph MeO (4.70)

65%

Conjugated dienes also give oxetanes, either by reaction with the singlet (n,π^*) excited state of aliphatic ketones (4.71), or by interaction with the triplet (n,π^*) state of aryl ketones (4.72). These processes need to be taken into consideration when dienes are used as triplet quenchers to study the photochemistry of carbonyl compounds—the singlet reaction means that some singlet quenching accompanies any observed triplet quenching, and the triplet reaction, though very inefficient, is in direct competition with triplet quenching. On a more useful note, aldehydes react with cyclic conjugated dienes to give good yields of bicyclic oxetanes, and that from propanal and cyclohexa-1,3-diene can be converted quite readily into non-6-en-1-ol (4.73), a pheromone of the Mediterranean fruit-fly.

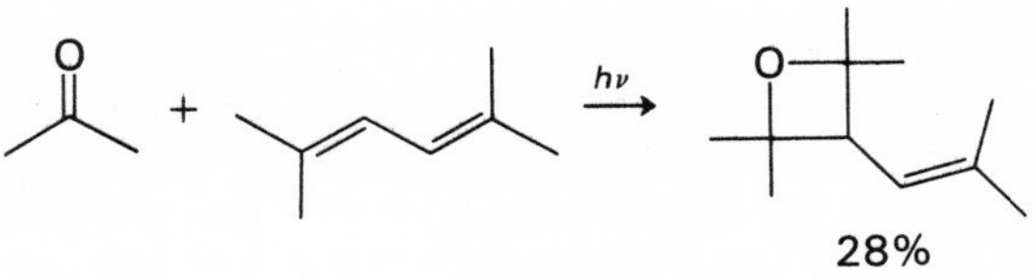

O + $\xrightarrow{h\nu}$ O 28% (4.71)

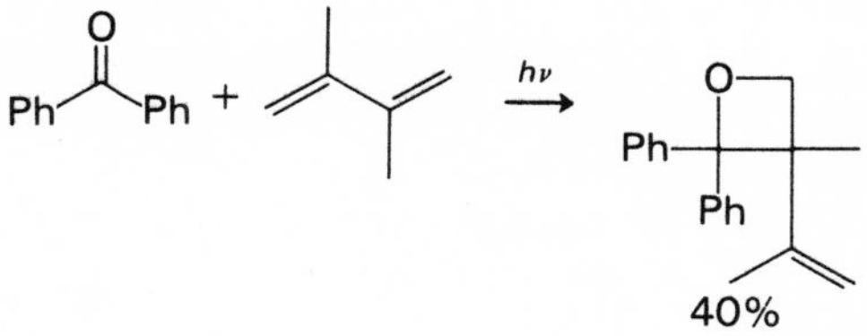

O Ph Ph + $\xrightarrow{h\nu}$ O Ph Ph 40% (4.72)

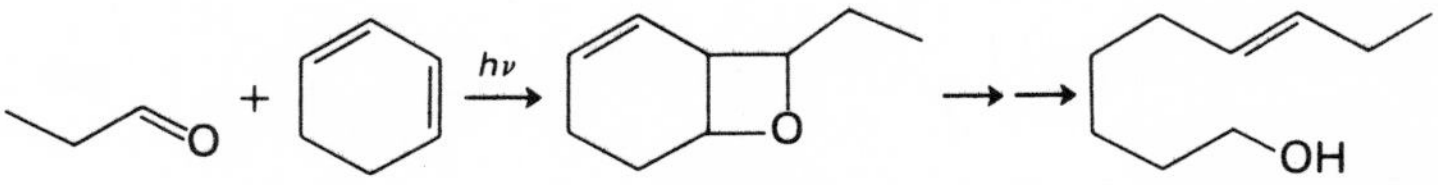

O + $\xrightarrow{h\nu}$ O 80% $\rightarrow\rightarrow$ OH (4.73)

Oxetanes can be formed by intramolecular reaction between a carbonyl group and an alkene, and this has been used (4.74) in making analogues of thromboxane A_2 (one of the compounds responsible for the control of blood clotting), albeit usually as the minor product. A special case of intramolecular reaction is seen for α,β-unsaturated carboxylic acids (4.75), where the product is an oxete that is tautomeric with a β-lactone. Oxetes may also be formed by photocycloaddition of ketones or aldehydes with alkynes; the oxete normally ring-opens at room temperature to give an α,β-unsaturated carbonyl compound (4.76), but at lower temperatures its spectral

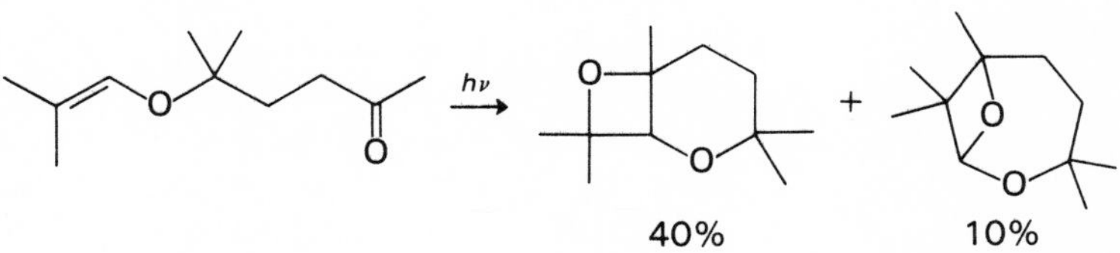

O O $\xrightarrow{h\nu}$ O O 40% + O O 10% (4.74)

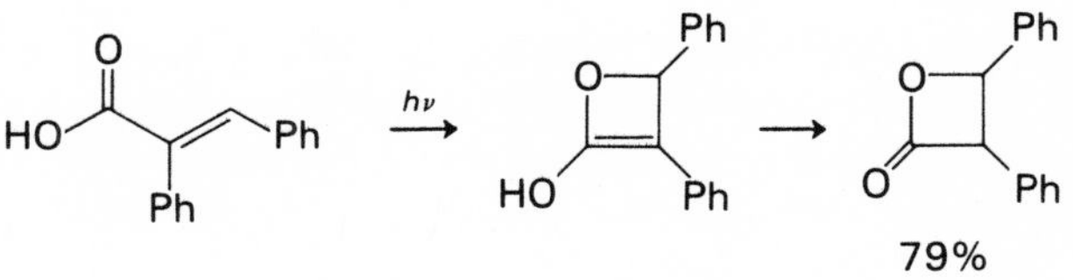

O HO Ph Ph $\xrightarrow{h\nu}$ O Ph HO Ph $\rightarrow$ O Ph O Ph 79% (4.75)

properties can be studied, and it may react further with a second molecule of excited carbonyl compound under these conditions (4.77).

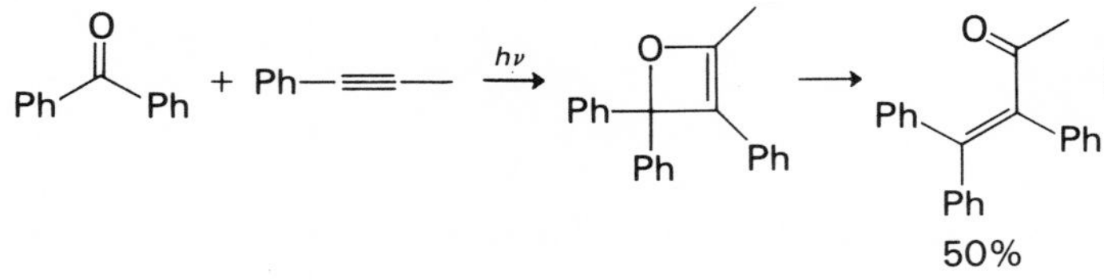

Ph, Ph, O; + Ph—≡—; $h\nu$; O, Ph, Ph, Ph; O, Ph, Ph, Ph; 50% (4.76)

O, Ph, H; + —≡—; $h\nu$, −78 °C; O, Ph, H; 20 °C; O, Ph; 43%; $h\nu$, PhCH=O; O, O, Ph, Ph (4.77)

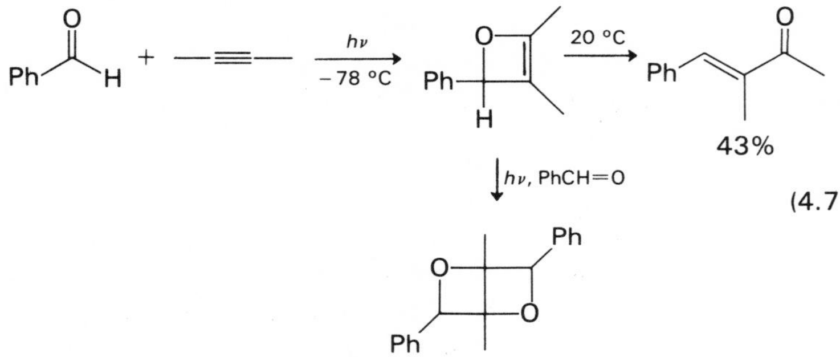

Aromatic dicarboximides behave in a different way on irradiation with alkenes. Instead of cycloaddition to the carbon–oxygen double bond, the overall reaction gives a product in which the alkene is inserted between the nitrogen atom and one of the carbonyl groups (4.78). The reaction is stereospecific, and the mechanism may

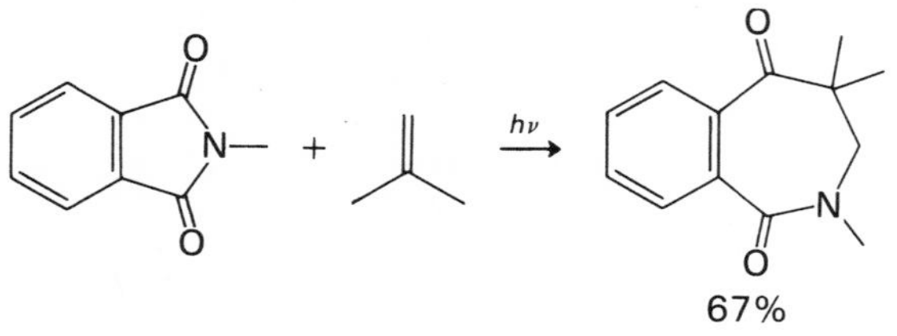

O, N—, O; + ; $h\nu$; O, N, O; 67% (4.78)

involve radical ions obtained by electron-transfer from the alkene to the excited imide. The intervention of ionic intermediates seems very likely in view of the alternative type of product formed when some alkenes are used in methanol solution as addends for *N*-methylphthalimide (4.79).

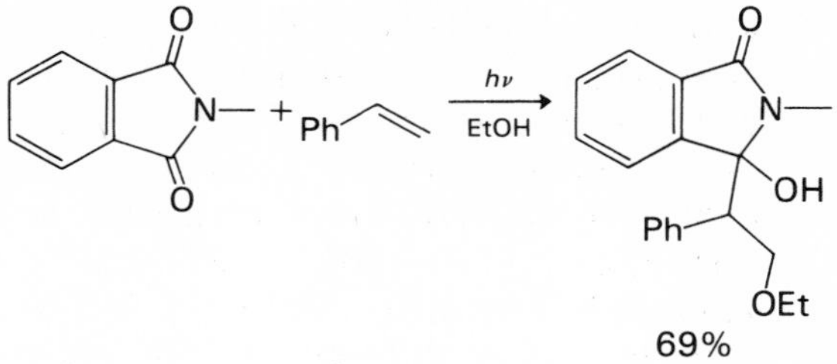

(4.79)

69%

Rearrangement of cyclohexenones and cyclohexadienones

Cyclic conjugated enones and dienones, especially those with a six-membered ring, undergo a number of photochemical rearrangements that set them apart from their acyclic analogues. Cyclohex-2-enone itself gives mainly cyclobutane-dimers by (2 + 2) photocycloaddition (see p. 65), and photoreduction occurs for some enones on irradiation in solvents such as propan-2-ol, but most cyclohex-2-enones carrying substituents at carbon-4 take part in the photorearrangements. Two basic types of reaction are observed, which involve migration of either the bond between C-4 and C-5 or of a substituent on C-4; in both cases the product is a bicyclo[3.1.0]hexan-2-one. The shift of the C-5 ring carbon is typical of 4,4-dialkylcyclohex-2-enones (4.80); C-4 becomes bonded to C-2, and C-5 to C-3, in the product. The reaction is common in bicyclic systems also, leading to fused tricyclic products (4.81).

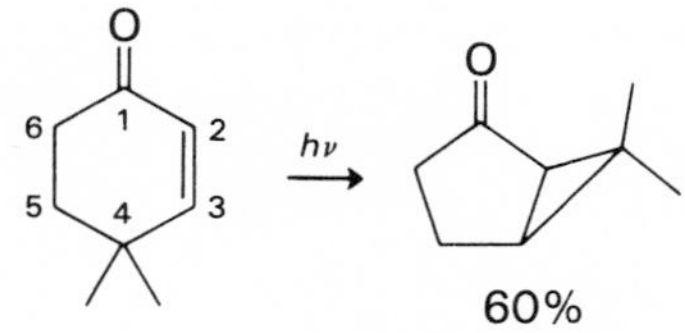

(4.80)

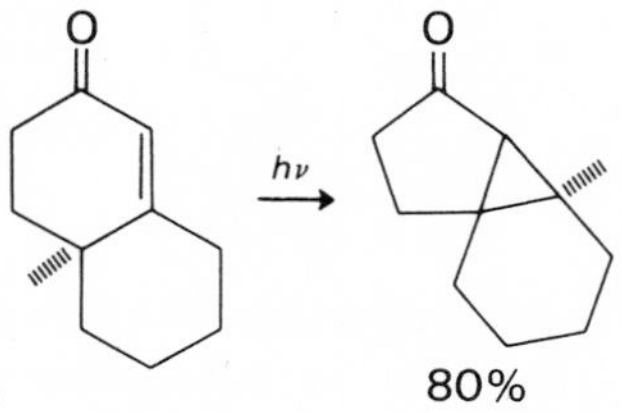

60%

(4.81)

80%

A mechanism involving a discrete zwitterionic intermediate appears to account for the diversion of the reaction to form monocyclic

clic products in aqueous acetic acid (4.82), but such an intermediate is inconsistent with the observed stereospecifity of the process, as evidenced by complete inversion of configuration at C-4 (4.83); this rules out the possibility of an intermediate that is achiral at C-4. Further mechanistic studies reveal that the reactive excited state is the (π,π^*) triplet state, which is surprising in the sense that completely stereospecific, apparently concerted, reactions are not common for triplet states. It seems likely that the reaction pathway involves a twisting about the carbon–carbon double bond in the triplet, which is related to the formation of transient *trans*-isomers for cyclohexenes (see p. 68). The twisted triplet decays largely to ground state, accounting for the very low quantum yields ($\phi < 0.01$), but a small proportion reacts to give the bicyclohexanone.

O O⁻ O
hν
+
HOAc
(4.82)
O O
+
OAc
20–25% 30–40%

O O
hν
aq. MeOH
(4.83)
Ph Ph

4.4-Diarylcyclohex-2-enones undergo a different photorearrangement to bicyclo[3.1.0]hexan-2-ones, in which an aryl substituent migrates from C-4 to C-3 (4.84). This reaction finds a parallel in the di-π-methane rearrangement of 3-phenylalkenes (see p. 54). It is usually efficient ($\phi = 0.1$–0.2), it occurs by way of the (n,π^*) triplet state, and there is a high degree of stereoselectivity. The parallel rearrangement of 4-alkyl-4-vinylcyclohex-2-enones (4.85) is like the di-π-methane reaction of 1,4-dienes. 4-Alkyl-4-arylcyclohex-2-enones can undergo either type of rearrangement, depending on the solvent; in hydroxylic solvents there is predominant migration of the C-5 ring carbon (4.83), whereas in benzene an aryl group migrates from C-4 (4.86). The effect of solvent is to alter the energies of the two low-lying triplet states, so that the (π,π*) state is lower in polar solvents, and the (*n*,π*) state in non-polar solvents.

O, Ph Ph $\xrightarrow{h\nu}$ O, Ph, Ph 81% (4.84)

O $\xrightarrow{h\nu}$ O 80% (4.85)

O, Ph $\xrightarrow[C_6H_6]{h\nu}$ O, Ph (4.86)

Irradiation of 4,4-disubstituted cyclohexa-2,5-dienones yields bicyclo[3.1.0]hex-3-en-2-ones (4.87) in a process that resembles the first type of rearrangement just described for cyclohexenones. The

O, Ph Ph $\xrightarrow{h\nu}$ O, Ph, Ph ~100% (4.87)

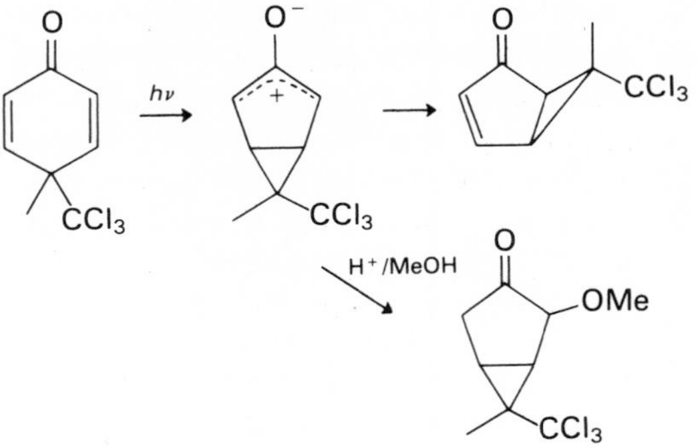

(4.88)

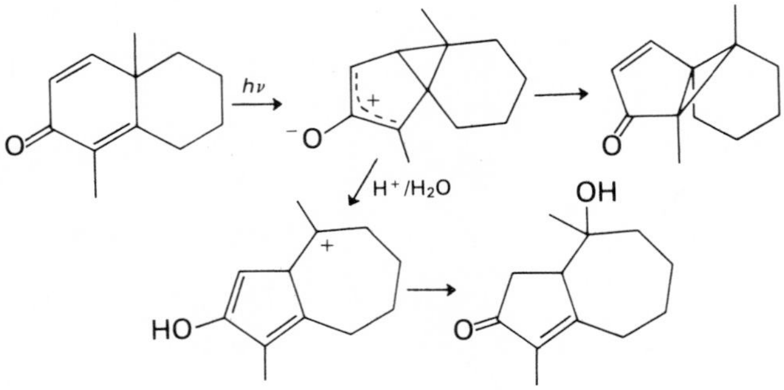

(4.89)

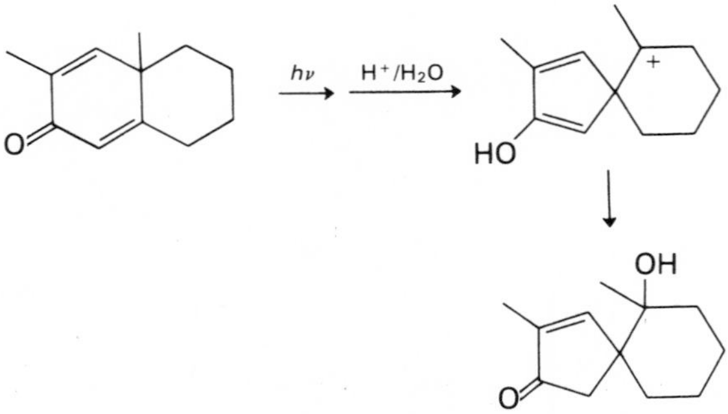

(4.90)

reaction occurs through the (n,π^*) triplet state and proceeds with inversion of configuration at C-4. However, the quantum yield is very high ($\phi = 0.8$–1.0), and the second double bond in the ring must play a significant role in the mechanism. A pathway involving bonding between C-3 and C-5 to give a zwitterionic intermediate has been proposed, and the formation in some cases of an adduct with methanol under acidic conditions (4.88) is in keeping with this suggestion. Related zwitterions have also been generated by alternative (non-photochemical) methods, and they give rise to the same bicyclohexenones. For bicyclic systems the use of an aqueous acidic medium diverts the reaction from the pathway leading to rearranged product, and gives instead hydrated products with a rearranged carbon skeleton (4.89). Such rearrangements, and the sensitivity to alkyl substitution (4.90), are characteristic of reactions involving carbonium ion intermediates.

The photoreactions of cyclohexa-2,5-dienones can give additional products under certain conditions, especially phenols. Sometimes these arise by a subsequent photoreaction of the bicyclohexenone (4.91), but they may also be produced by photoreduction if a hydrogen-donor solvent is employed and a good radical leaving group is present at C-4 (4.92).

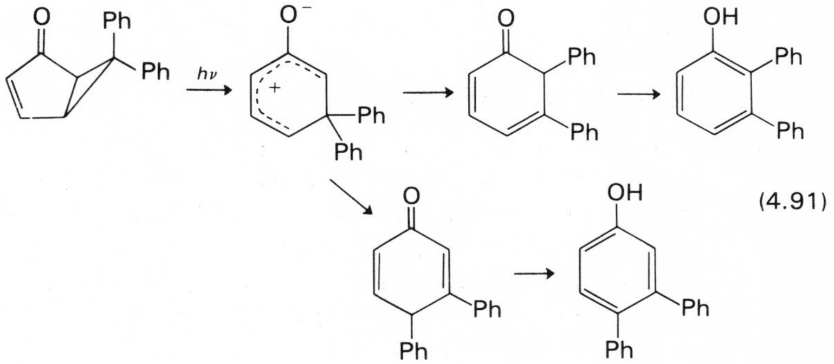

(4.91)

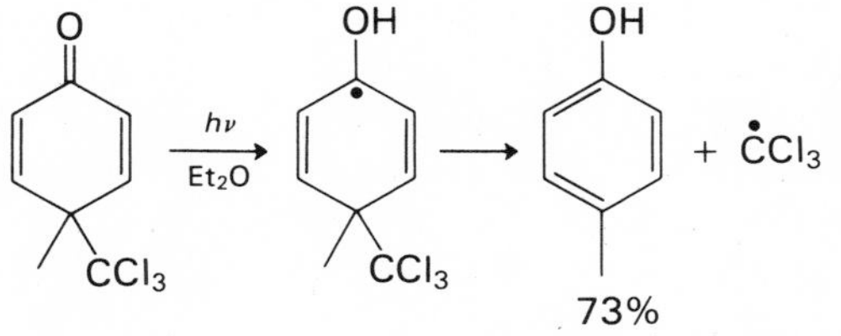

(4.92)

Finally in this section, we look at the isomeric cyclohexa-2,4-dienones. On irradiation these undergo ring-opening to give an unsaturated ketene (4.93), which can be detected spectroscopically if the reaction is carried out at low temperature. In the presence of a good nucleophile such as an amine, the ketene gives good yields of a dienoic acid derivative. In the absence of such a nucleophile, the ketene either reverts to the original cyclohexa-2,4-dienone, or it

$h\nu$ $C_6H_{11}NH_2$ NHC_6H_{11} 90% (4.93)

cyclizes in a different manner to give a bicyclo[3.1.0]hex-3-en-2-one. There is evidence that for some systems the bicyclohexenone may also be formed directly from an excited state of the dienone.

Thiocarbonyl compounds

The sulfur analogues of ketones are thioketones, containing the C=S functional group, and they offer an interesting array of photochemistry that contrasts in many ways with that of ketones. Thioketones are deeply coloured, which implies that they absorb in the visible range of wavelengths; an $n \rightarrow \pi^*$ absorption band is largely responsible for the colour, and the energy of the (n,π^*) excited states is considerably lower than for the corresponding ketones. One consequence of this lower energy is that α-cleavage reactions are relatively unimportant in the thiocarbonyl series. The energy difference between S_1 (n,π^*) and S_2 (π,π^*) for thioketones is very large, and irradiation of thioketones with ultraviolet radiation often leads to reactions that occur, quite unusually, through the upper excited singlet state. Conversely, the energy difference between S_1 (n,π^*) and T_1 (n,π^*) is very small, so intersystem crossing is very fast, and irradiation with visible light usually leads to reaction through the lowest triplet state. The reactions of the S_2 and T_1 states are sometimes quite different.

S $\xrightarrow[PhCH_2OH]{h\nu}$ SH (4.94)

82%

Thioketones are readily photoreduced (4.94), but in many cases a variety of products is formed rather than just one major product. This is because the intermediate radicals containing sulfur atoms undergo a wider range of reactions than do related radicals containing oxygen. Intramolecular hydrogen abstraction occurs efficiently to give cyclization products. In an unconstrained system there is a preference for abstraction from the δ-position (4.95) if the reactive excited state is S_2; when γ-abstraction does take place (4.96), it occurs through the T_1 triplet state. If no other suitable hydrogen atom is available, β-abstraction to give a cyclopropanethiol (4.97) occurs quite readily.

S Ph hν >300 nm SH Ph 70% (4.95)

S Ph O hν >445 nm SH Ph O 45% (4.96)

S Ph hν Ph HS 82% (4.97)

When thioketones are irradiated alone, a cyclodimerization may occur to give a 1,3-dithietane (4.98). In the presence of an alkene, different cycloadducts are found, usually thietanes. If visible radiation is used, electron-rich alkenes are especially effective as addends, and the products can be rationalized on the basis of a two-step mechanism involving the more highly stabilized biradical intermediate (4.99). Sometimes a 1,4-dithiane accompanies the thietane (4.100) as a result of the trapping of the biradical by a further molecule of ground-state thioketones (unlike ketones, thioketones react with radicals quite readily, which is one of the causes of

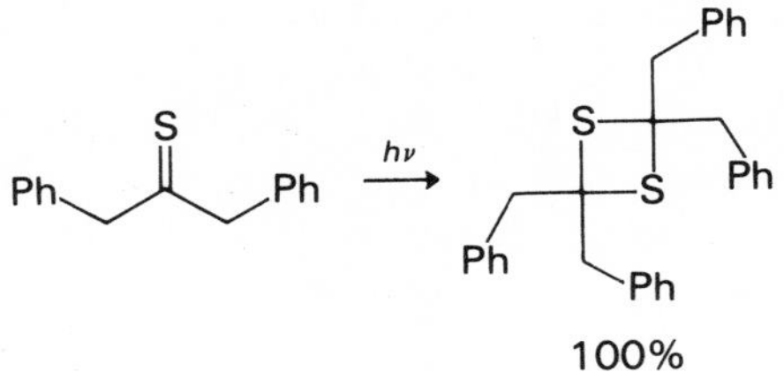

(4.98)

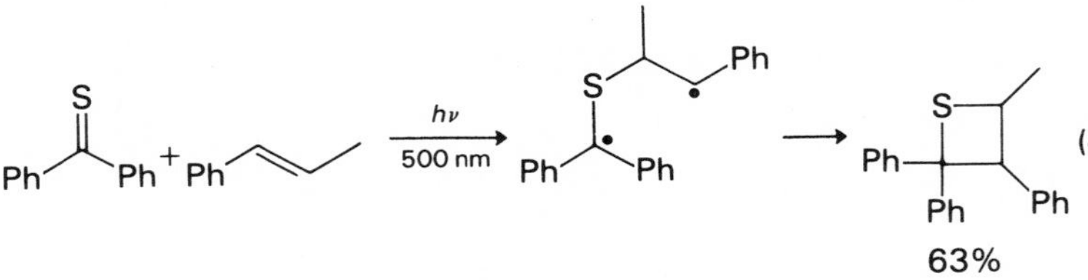

100%

(4.99)

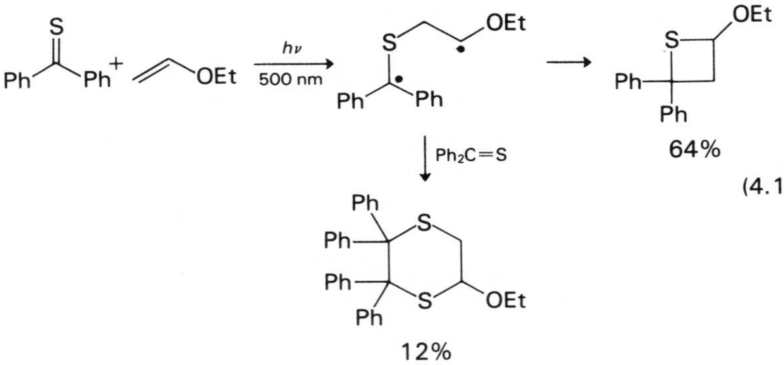

63%

(4.100)

64%

12%

complications in the photoreduction reactions of thiocarbonyl compounds).

When shorter (ultraviolet) wavelengths are used, photocycloaddition often occurs through the S_2 (π,π^*) state rather than through the lowest triplet state. Thietanes are again the principal products, but they can be formed with most classes of alkene, including electron-deficient alkenes (4.101) that are not very successful when used with 500 nm light, and they are formed in a highly stereoselective manner. With alkynes as addends, thietes are the major products from most thiocarbonyl compounds (4.102), and unlike their oxygen analogues (oxetes) they are stable at room temperature. In some cases diversion of the biradical intermediate by attack on a neighbouring aromatic ring provides a route to an alternative major product (4.103).

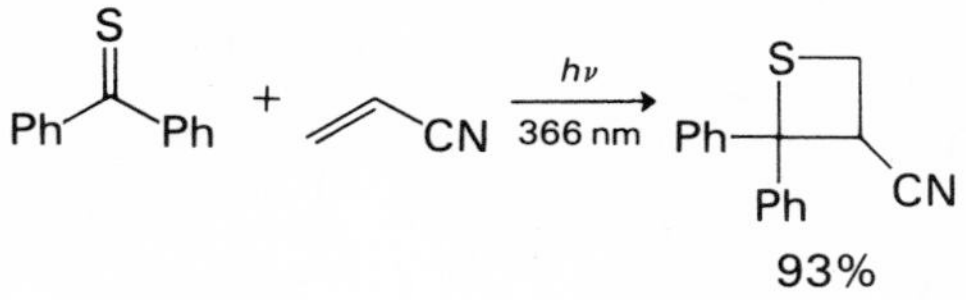

(4.101)

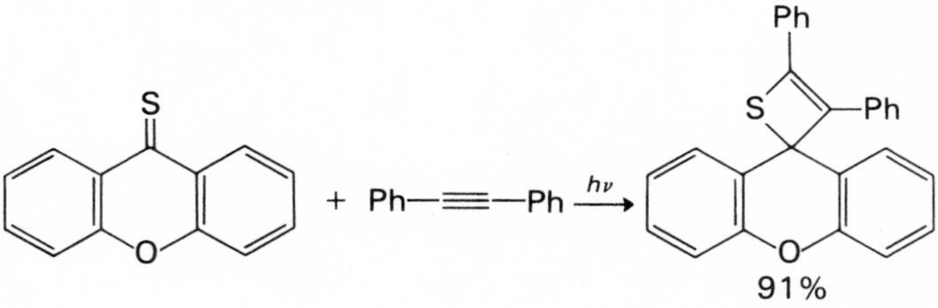

93%

(4.102)

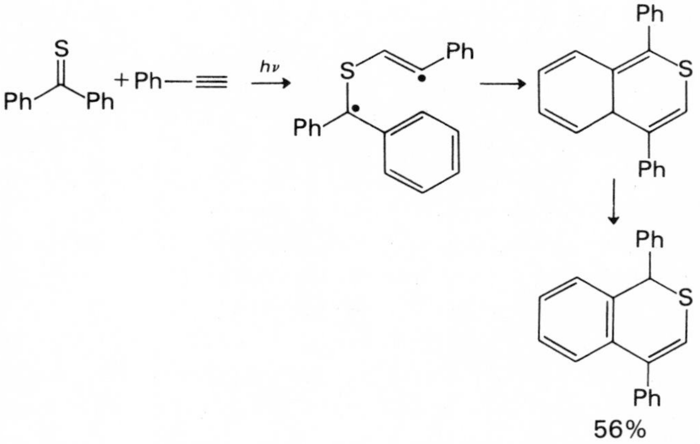

91%

(4.103)

56%

Two types of reaction are observed for thioketones that do not have parallels in ketone photochemistry. The first is photocyclization in thioketones with polycyclic aromatic groups (4.104), where the sulfur atom forms a new bond to the aromatic system. The second is photo-oxidation by singlet oxygen to give the corresponding carbonyl compound (4.105), possibly via a 1,2,3-dioxathietane formed as a result of initial cycloaddition.

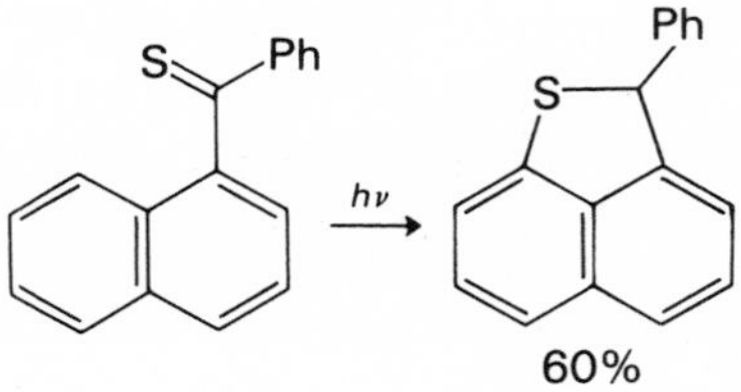

(4.104)

60%

$h\nu$, O_2 / methylene blue

100%

(4.105)

Further reading

D. S. Weiss, in A. Padwa (ed.), *Organic Photochemistry*, vol. 5, Dekker (1981). This detailed review covers the photochemical alpha-cleavage reactions of cyclic ketones.

P. J. Wagner, in P. de Mayo (ed.), *Rearrangements in Ground and Excited States*, vol. 3, Academic Press (1980). The photorearrangements of a simple carbonyl compounds that occur by way of biradicals are the subject of this account, including reactions that involve intramolecular hydrogen abstraction.

G. Jones, in A. Padwa (ed.), *Organic Photochemistry*, vol. 5, Dekker (1981). This is a review of synthetic applications of the photochemical cycloaddition of carbonyl compounds.

H. A. J. Carless, in W. M. Horspool (ed.), *Synthetic Organic Photochemistry*, Plenum (1984). The photochemical synthesis of oxetans is discussed, with both synthetic and mechanistic aspects highlighted.

D. I. Schuster, in P. de Mayo (ed.), *Rearrangements in Ground and Excited States*, vol. 3, Academic Press (1980); K. Schaffner and M. Demuth, in the same volume. These reviews cover the photochemical rearrangements of enones (especially cyclic enones) and of conjugated cyclic dienones, respectively.

P. H. Mazzocchi, in A. Padwa (ed.), *Organic Photochemistry*, vol. 5, Dekker (1981). The photochemistry of imides is described, which has proved to be very versatile and in some respects different from that of other carbonyl compounds.

J.D. Coyle, *Tetrahedron*, vol. 41, Pergamon (1985), p. 5393. This is a review of the photochemistry of thiocarbonyl compounds, the sulfur analogues of carbonyl compounds.

CHAPTER 5

Photochemistry of other organic compounds

In the previous three chapters we have looked at the photochemistry of alkenes, substituted benzenes, ketones and related compounds. These classes of organic compound are important and interesting whether studied in their excited state (photochemistry) or in their ground state (thermal chemistry). However, when it comes to choosing from the wide range of other functional groups for the purpose of an introductory survey such as this, those compounds chosen because of their photochemical interest are not always the same as those that might be chosen for an account of their thermal chemistry. There are two main reasons for this. First, a prerequisite for photochemical reaction is that the compound absorbs visible or ultraviolet light, or that it can be sensitized by another light-absorbing compound, to produce an electronically excited state. Many saturated compounds that play a major role in thermal chemistry, such as alcohols or amines, do not absorb appreciably at wavelengths longer than about 220 nm, and their photochemistry cannot be studied as conveniently as that of compounds absorbing at longer wavelengths for which the practical requirements are much less stringent. The second reason is related to this, because absorption of high-energy ultraviolet radiation by saturated compounds generally leads to electronically excited states that dissociate efficiently; the overall reaction is often a fragmentation that is not 'useful' from an organic chemist's point of view, although there are interesting theoretical aspects of such fragmentations. It is for this reason that haloalkanes (alkyl halides) play only a minor role in photochemistry as compared with their important place in thermal chemistry.

These considerations lie behind the choice of compounds for the survey in this final chapter. It includes those with multiply bonded carbon–nitrogen groups (e.g. imines, nitriles), compounds with functional groups incorporating two or three nitrogen atoms (azo- or diazo-compounds, azides), compounds with groups containing nitrogen and oxygen atoms (nitrites, nitro-compounds), those with oxygen or sulfur chromophores (e.g. peroxides, thiols), and those with halogen chromophores (e.g. halides, hypohalites). Also included are a number of important photochemical reactions involving alkanes, in which the role of the light is to generate a species capable of attacking alkanes (as in halogenation or nitrosation).

Imines, iminium salts and nitriles

The electronic structure of imines ($R_2C{=}NR$) is similar to that of ketones ($R_2C{=}O$), and imines have ultraviolet absorption characteristics that resemble those of their oxygen counterparts. However, the longest-wavelength ($n \rightarrow \pi^*$) absorption maximum is generally at much shorter wavelength than that of analogous ketones; for example, saturated aliphatic imines have a weak (ϵ = 100–200 l mol^{-1} cm^{-1}) maximum at around 240 nm, compared with one around 280 nm for the $n \rightarrow \pi^*$ band of saturated ketones. Many simple imines are not easy to study because they hydrolyse very readily, and traces of moisture can lead to small amounts of ketone; the ketone is a stronger absorber and its excited state can take part in or initiate reactions that are not characteristic of the excited state of the imine. Because of this, more work has been carried out using derivatives that are less susceptible to hydrolysis, such as oximes ($R_2C{=}NOH$) or hydrazones ($R_2C{=}NNH_2$) but for many of these there is no evidence for a low-lying (n,π^*) excited state—instead, the ultraviolet absorption spectrum shows intense absorption characteristic of a $\pi \rightarrow \pi^*$ band.

In one respect >C=N– compounds are different from their oxygen analogues, because they can exist in two geometrically isomeric forms if the two groups attached to carbon are different. There is a greater resemblance in this feature to alkenes (see p. 42) or to azo-compounds (see p. 148), and a major photochemical reaction for many compounds is *E–Z* interconversion (the names *anti* and *syn* are sometimes used for geometrical isomers in this context). This is seen for the *O*-methyl derivative of acetophenone oxime (5.1); in the thermal equilibrium the predominant isomer (98%) is the *E*-compound, but on direct irradiation a *Z* : *E* ratio of 2.2 can be achieved. This ratio is governed largely by the relative absorption coefficients of the two isomers at the wavelength used for irradiation.

Triplet-sensitized isomerization can also be achieved, in which case the relative values of triplet energies are the most important factor.

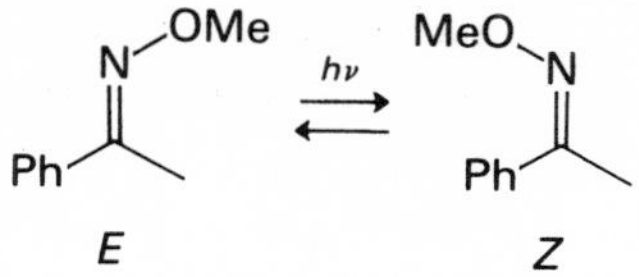

(5.1)

Photochemical cleavage can occur in compounds such as oximes or hydrazones, but it invariably involves dissociation of the nitrogen–oxygen or nitrogen–nitrogen bond. This can lead to efficient production of azines (5.2) as a result of dimerization of the nitrogen-centred radicals. Cleavage alpha to the C=N group, analogous to alpha-cleavage in ketones (see p. 107), is found only rarely. A group of compounds for which such a process is normal are the 2*H*-azirines; photochemical cleavage in this system provides a 1,3-dipolar species (5.3), which in the absence of an added reagent can attack unchanged azirine to give a bicyclic product, but which reacts with

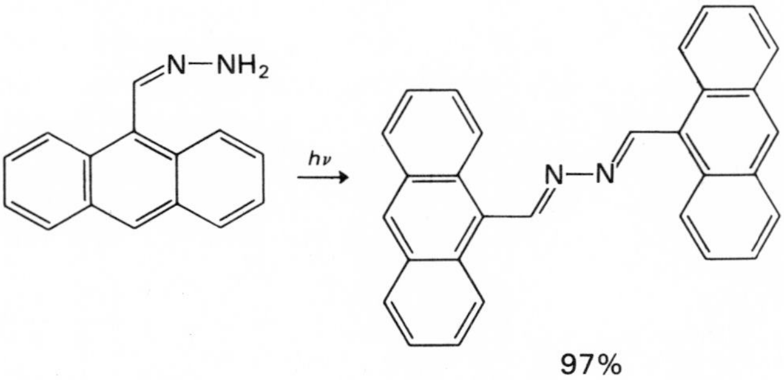

(5.2)

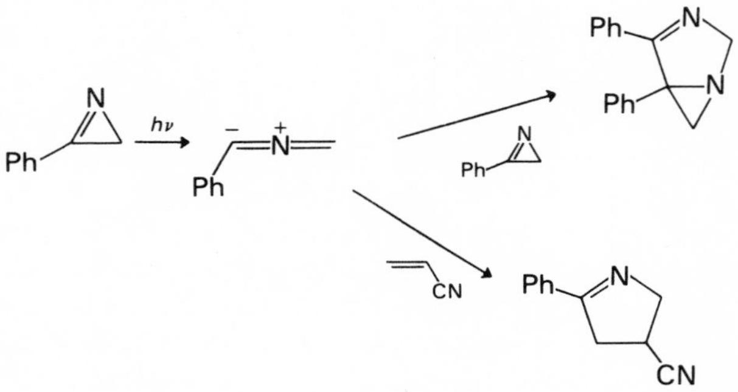

97%

(5.3)

dipolarophiles such as acrylonitrile to give a pyrrole derivative. The dipolar intermediate is reactive enough to combine even with carbon dioxide, leading to an oxazolinone (5.4). A parallel reaction is seen with aziridines, the saturated ring compounds related to azirines, and this again provides a useful route to a variety of pyrrole compounds (5.5).

Ph N; $h\nu$, CO_2; Ph N O O; 90% (5.4)

Ar N+ COOMe COOMe; MeOOC—≡—COOMe; Ar N MeOOC COOMe MeOOC COOMe; $h\nu$; Ar N MeOOC COOMe (5.5)

Imines can be photoreduced by hydrogen donors such as propan-2-ol, but in some cases it is clear that a small amount of carbonyl compound (arising from hydrolysis of the imine) is necessary to initiate the process. For example, *N*-alkylimines of benzaldehyde give ethane-1,2-diamine derivatives on irradiation in 95% ethanol (5.6), but they are inert when irradiated in a perfectly dry alcohol. The C=N chromophore is capable of abstracting a hydrogen atom, and both the photoreduction of a tetrahydropyridine (5.7) and the photo-addition of *p*-xylene to an isoxazoline (5.8) occur by such a direct reaction.

N R Ph; $h\nu$, 95% EtOH; NHR Ph Ph NHR; 95% (5.6)

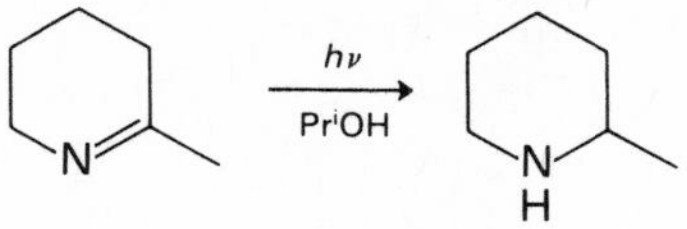

(5.7)

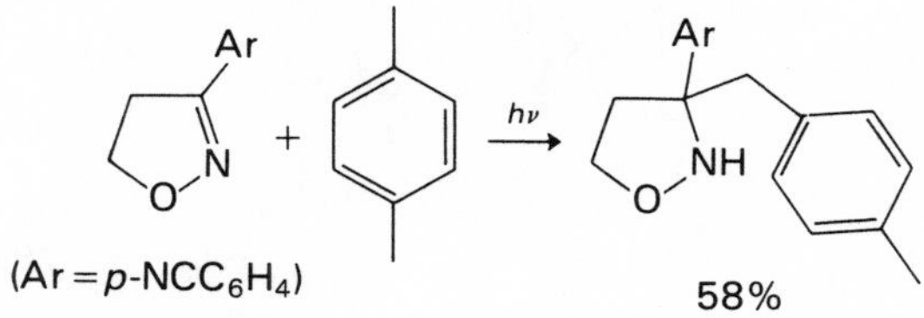

(5.8)

$(Ar = p\text{-}NCC_6H_4)$ 58%

Intramolecular hydrogen-transfer to an excited imine is responsible for the photochromism of derivatives of salicylaldehyde (*o*-hydroxybenzaldehyde); irradiation converts the colourless or yellow imine into a much more intensely coloured photoproduct (5.9), but removal of the light source allows rapid reversal of the process. Reactions of this kind form the basis of some organic photochromic glass formulations; exposure to increasingly bright light alters the proportion of the isomers in favour of the dark-coloured photoproduct.

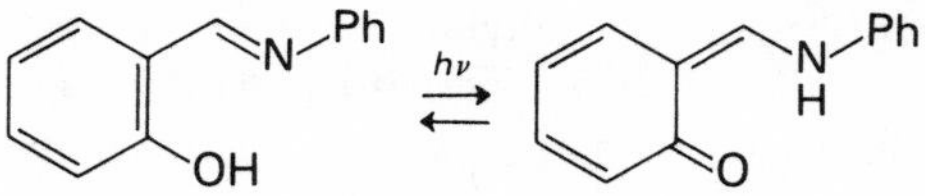

(5.9)

Like carbonyl compounds, those containing the C=N group can undergo (2 + 2) photocycloaddition with alkenes, and the product structures include four-membered nitrogen-heterocyclic rings. However, only a few examples have been reported, largely where the C=N bond is itself part of a five- or six-membered ring. For example, 4,5-dihydro-oxazol-4-ones react with alkenes to give bicyclic azetidines (5.10), and 1,4-benzoxazin-2-ones behave simi-

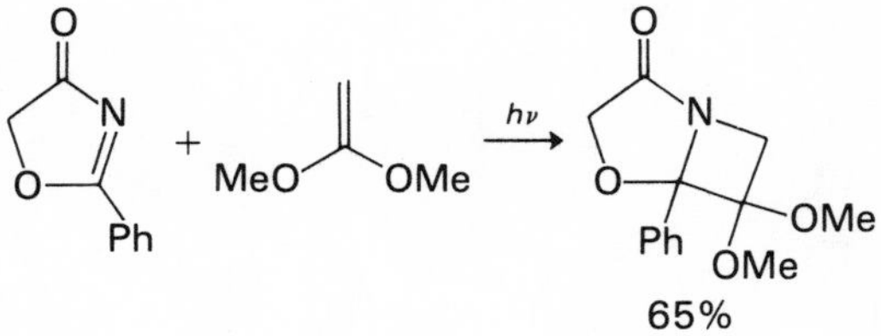

(5.10)

65%

larly with electron-deficient alkenes (5.11). In the absence of added alkene, irradiation of 2-phenylbenzoxazole leads to a dimer with a 1,3-diazetidine unit (5.12).

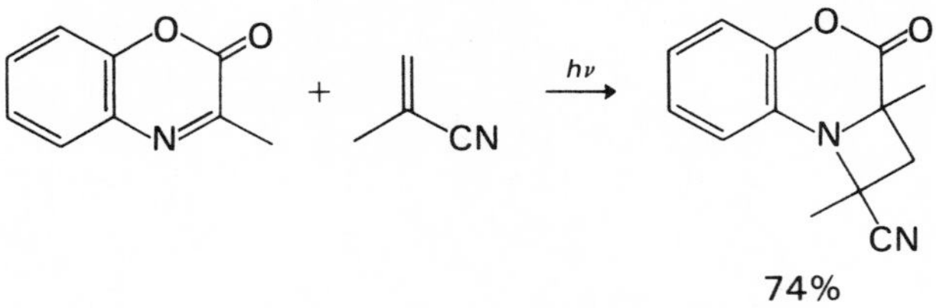

(5.11)

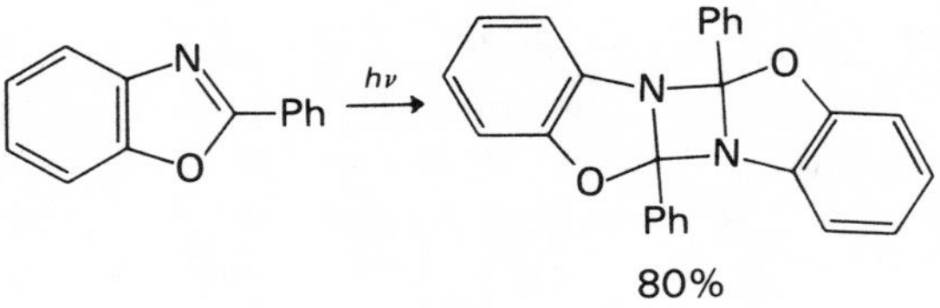

74%

(5.12)

80%

Imine photochemistry is rather overshadowed by the very much more extensive array of photochemical reactions known for ketones and related compounds, but an outstanding feature of the excited-state reactions of the nitrogen-containing compounds is the reactivity of protonated or alkylated imines. Iminium salts ($R_2C{=}\overset{+}{N}HR\ X^-$ or $R_2C{=}\overset{+}{N}R_2\ X^-$), especially those with aryl-conjugation, are well suited to photochemical reaction because their excited states are very good electron-acceptors. It is significant that the visual pigment rhodopsin contains an iminium linkage (5.13) at the point where the retinal chromophore is linked covalently to the protein (opsin). On irradiation with an alkene in methanol as solvent, an iminium salt gives rise to an adduct (5.14) that resembles those formed from imides under similar conditions (see p. 131), and a similar electron-transfer mechanism seems likely. In intramolecular versions of the reaction, 1-azabicycloalkanes can be formed (5.15), and such ring systems are

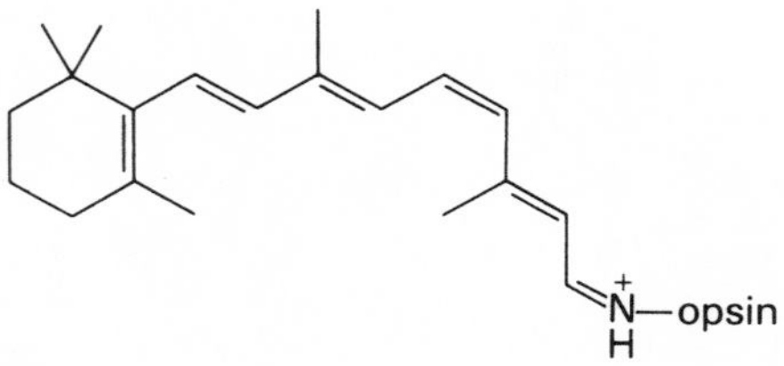

(5.13)

present in the group of natural products known as the pyrrolizidine or quinolizidine alkaloids (containing bicyclo[3.3.0] or bicyclo[4.3.0] frameworks, respectively).

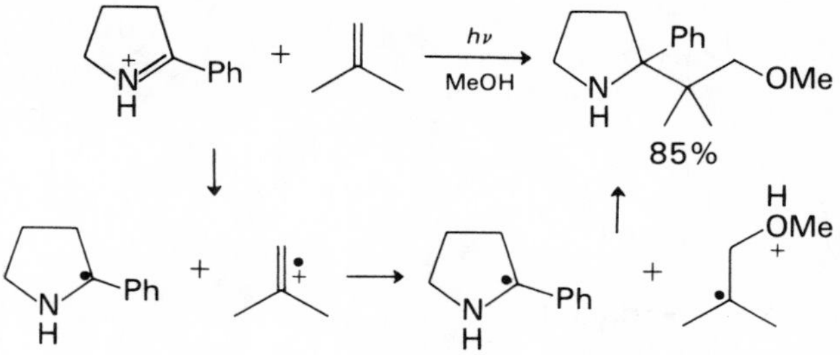

(5.14)

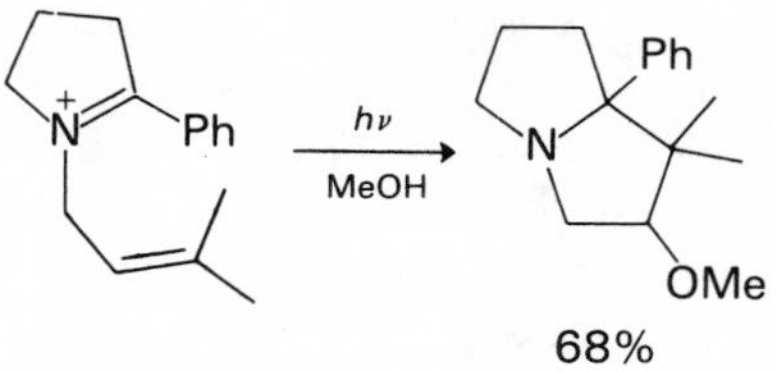

(5.15)

Irradiation of an iminium salt with toluene (or with benzyltrimethylsilane) yields a benzyl-substituted amine (5.16), and this reaction also can be adapted to produce multicyclic nitrogen heterocycles by intramolecular reaction.

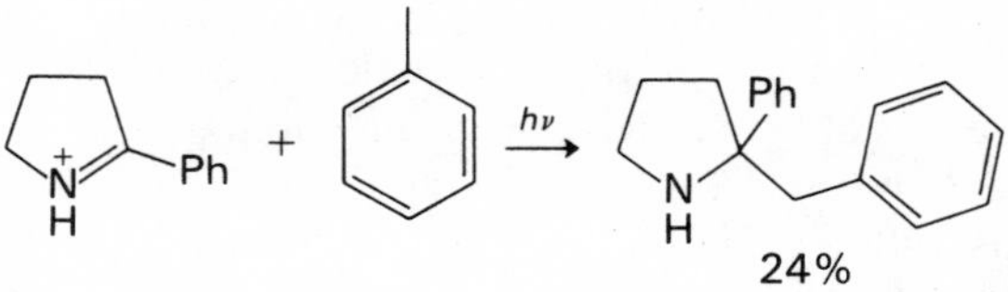

(5.16)

Aliphatic nitriles (RC≡N) absorb only in the short-wavelength ultraviolet region (λ_{max}~170 nm); acetonitrile (MeCN) is a good solvent for many photochemical reactions, and its high polarity coupled with an absence of hydroxyl groups makes it especially useful for those processes that go by way of an electron-transfer mechanism. Aromatic nitriles (ArC≡N) are more amenable to photochemical study, though they turn out to be relatively inert as far as the C≡N group is concerned. Benzonitrile does undergo photocycloaddition with electron-rich alkenes (5.17); the products are a four-membered cyclic imine derived by (2 + 2) cycloaddition, and a compound related to it by subsequent electrocyclic ring-opening.

$$Ph{-}C{\equiv}N \; + \; \text{(alkene)} \xrightarrow{h\nu} \text{(azetine, 8\%)} \rightleftarrows \text{(2-azadiene, 66\%)} \qquad (5.17)$$

Azo-compounds

Azo-alkanes (RN=NR) are yellow; their electronic absorption spectra show a low-intensity maximum in the near ultraviolet that can be attributed to an $n \rightarrow \pi^*$ transition. The low energy of this transition arises because the in-plane overlap of non-bonding orbitals on the two nitrogen atoms leads to one being of quite high energy. The same rationalization accounts for the observation that *cis* isomers (λ_{max}~380 nm, ϵ_{max} 150–200) are more strongly coloured than *trans* isomers (λ_{max} ~350 nm, ϵ_{max}15–20). Simple aromatic azo-compounds such as azobenzene (PhN=NPh) are orange because the $n \rightarrow \pi^*$ absorption maximum is in the visible wavelength range (λ_{max} ~440 nm). Substituted azobenzenes, or azo-compounds with different aromatic groups such as substituted naphthalenes, can be very highly coloured, and such azo dyes or pigments form one of the largest group of commercial organic colouring materials. The photochemistry of azo-compounds is therefore of importance, since in most applications a high degree of 'light-fastness' (stability towards photochemical change) is required, such that irreversible change occurs with a quantum yield much lower than 0.0001.

Aliphatic azo-compounds undergo *cis–trans* isomerization on irradiation, and in many instances the less stable isomer can be isolated in reasonable yield after irradiating the more stable one with light of a suitable wavelength (remember that the ratio of isomers at the photostationary state depends largely on the outcome of the competition for available light by the two isomers). Thus acyclic *cis* azoalkanes (5.18) and cyclic *trans* azoalkanes (5.19) can be prepared photochemically. The mechanism of the isomerization may involve twisting about the central (N—N) bond in the excited state (as for the corresponding reaction of alkenes), or inversion at one of the

(5.18)

40% 60% (5.19)

nitrogen atoms may be responsible, but the observable outcome rarely indicates such mechanistic detail. A competing reaction for some of the simplest azoalkanes, such as azomethane (MeN=NMe) is that the *cis* isomer is not stable at room temperature, but breaks down to give molecular nitrogen and two alkyl radicals.

Most *cis*-azoalkanes do this on heating, or on irradiation with shorter-wavelength radiation, and photolysis of azoalkanes provides a convenient source of certain radicals; the photochemical process can be more readily controlled than the thermal reaction. Radicals produced in this way may be employed as initiators for the polymerization of alkenes, and a widely used compound for this purpose is azobisisobutyronitrile (ABIN, 5.20).

$$\xrightarrow{h\nu} N_2 + 2Me_2\dot{C}{-}CN \qquad (5.20)$$

If nitrogen is eliminated from a cyclic azoalkane by irradiation, the initially formed biradical may react by internal combination to form a smaller-ring cycloalkane. On direct irradiation this is often a highly stereoselective process (5.21), although mixtures of isomers are generally formed when triplet sensitization operates. This reaction

hν, PhCOMe — 90% + 10% (5.22)

COOMe → hν → COOMe, 87% (5.21)

provides a route to highly strained compounds such as bicyclo[2.1.0]pentane (5.22) or methylenecyclopropane (5.23), and it is widely employed in routes to quite exotic cage-type molecules (5.24). Normally the use of light rather than heat to promote bond-cleavage leads to a much higher yield of the required cyclic compound rather than ring-opened products.

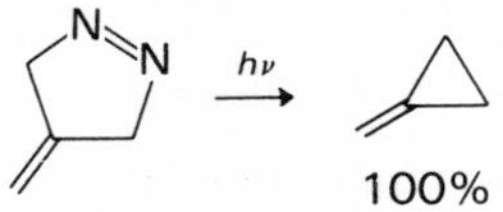

N N $h\nu$ 100% (5.23)

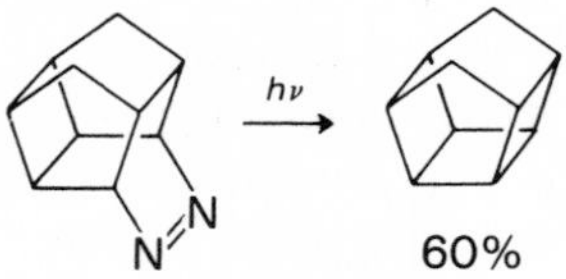

N N $h\nu$ 60% (5.24)

Aromatic azo-compounds also undergo photochemical *cis–trans* isomerization (5.25), but cleavage to give aryl radicals is not an important process; this is because the C—N bond strength is higher, and the excited-state energy lower, than for the aliphatic counterparts. Such isomerization accounts for slight, reversible changes of

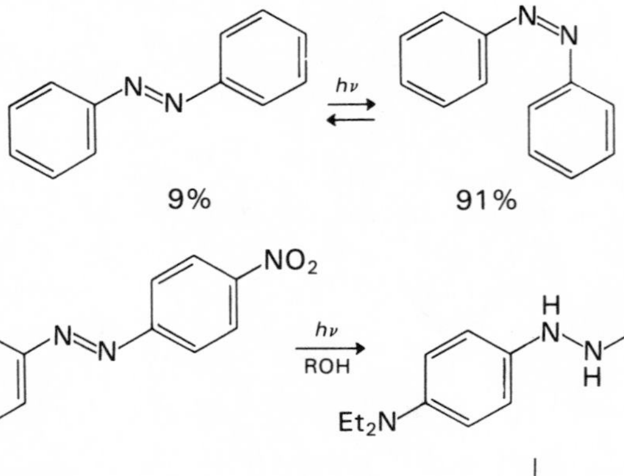

N N $h\nu$ N N 9% 91% (5.25)

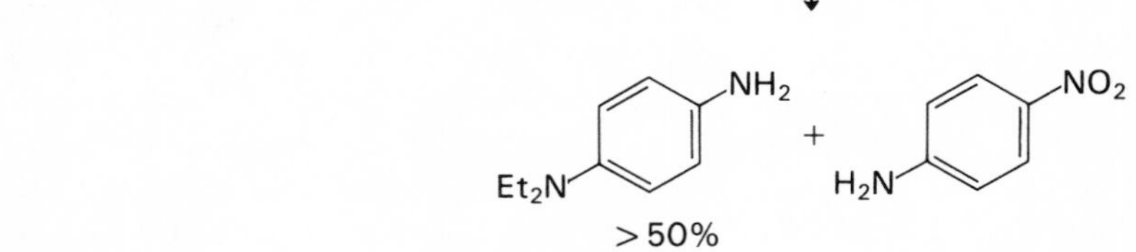

Et_2N N N NO_2 $h\nu$ ROH Et_2N H N N H NO_2 (5.26)

Et_2N NH_2 + H_2N NO_2

>50%

shade in some azo-dyes, but the effect is often noticeable only to the trained colourist. More permanent fading is a problem with certain types of azo-dye, and although the mechanisms are not fully elucidated it seems that photoreduction or photo-oxidation can be responsible for this. Azobenzene is photoreduced to hydrazobenzene (PhNHNHPh) by irradiation in propan-2-ol; other azo-compounds are similarly reduced, and the process may go further (5.26) to yield much less highly coloured amines.

The N=N double bond does take part in a few photocycloaddition reactions to give cyclic compounds with two adjacent nitrogen atoms in the ring. Intermolecular (2 + 2) cycloadditions are not known, but some intramolecular examples of this reaction are reported for quite complex compounds (5.27) in which the reacting groups are held fairly rigidly in an orientation suitable for reaction. A (4 + 2) cycloaddition takes place when naphthalene is irradiated with an electron-deficient cyclic azo-compound (5.28).

$h\nu$ 88% (5.27)

$h\nu$ 40% (5.28)

Diazo-compounds, diazonium salts and azides

Diazo-compounds ($R_2C{=}N_2$) are coloured yellow on account of an $n \rightarrow \pi^*$ absorption band in the violet-blue region of the visible spectrum (λ_{max} 400–500 nm). They lose molecular nitrogen on irradiation, to give divalent carbon species known as carbenes ($R_2\ddot{C}$); direct irradiation leads to a singlet carbene (the two electrons are spin-paired), which may undergo intersystem crossing to the lower-energy triplet carbene (the two electrons have parallel spin); the triplet carbene is formed directly from the diazo-compound if a triplet sensitizer is employed.

The diverse chemistry of carbenes is beyond the scope of this account, but a few typical reactions are shown here to illustrate the usefulness of the photochemical generation of these reactive species. A carbene can insert into a C–H bond, and this finds application in the reaction of an α-diazoamide to produce a β-lactam (5.29). Carbenes derived from α-diazoketones can rearrange to ketenes, and thus a route is opened up to ring-contraction for making more highly strained systems (5.30). Carbenes also react with alkenes, often by cycloaddition to yield cyclopropanes in a process that can be very efficient (5.31) and highly stereoselective (5.32).

80% (5.29)

60% (5.30)

100% (5.31)

(5.32)

Occasionally a diazo-compound on irradiation isomerizes to a diazirine (5.33); although this process could in principle precede the loss of nitrogen in any diazo system, there is no evidence to suggest that such isomerization occurs in most cases.

97%

N_2 ... F_3C ... CF_3 ... O $\xrightarrow{h\nu}$ N=N ... F_3C ... CF_3 ... O (5.33)

Organic azides (RN_3) are isoelectronic with diazo-compounds, and like them are yellow in colour. On irradiation azides lose nitrogen to produce monovalent nitrogen species, nitrenes (RN). In the absence of an addend, aliphatic nitrenes generally undergo a shift of hydrogen to give an imine (5.34), whereas aromatic nitrenes can dimerize to yield an azo-compound (5.35).

N_3 $\xrightarrow{h\nu}\rightarrow$ NH (5.34)

70%

MeO ... N_3 $\xrightarrow{h\nu}\rightarrow$ OMe ... N=N ... MeO (5.35)

82%

The dimerization appears to be a characteristic of triplet nitrenes, since it occurs on sensitization by such compounds as benzophenone. When 2-azidobiphenyl is irradiated in the presence of a triplet quencher (penta-1,3-diene) to inhibit the formation of an azo-compound, the main product is carbazole (5.36), which is formed by internal insertion into an aromatic C−H bond. Other nitrenes may also participate in intramolecular insertion, and this allows lactams to be formed from acyl azides (5.37).

N_3 $\xrightarrow{h\nu}\rightarrow$ N H (5.36)

89%

N_3 O $\xrightarrow{h\nu}\rightarrow$ NH O (5.37)

Photogenerated nitrenes can undergo cycloaddition with alkenes; intermolecular reaction leads to aziridine products (5.38), and intramolecular reaction in vinyl azides gives azirines (5.39). The bicyclic azirine from phenyl azide has not been isolated, but it is the intermediate that best accounts for the formation of a substituted azepine when this azide is irradiated in the presence of a secondary amine (5.40).

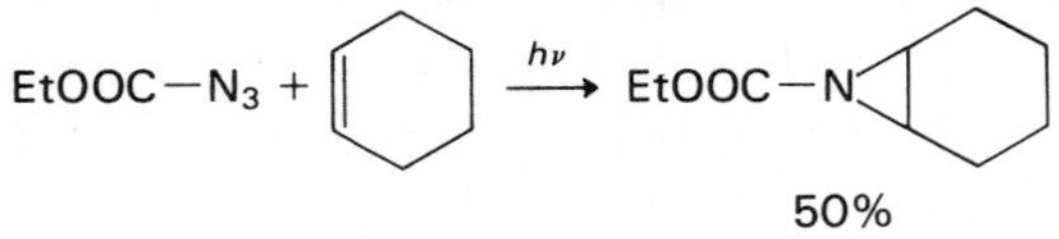

$$\mathrm{EtOOC{-}N_3} + \xrightarrow{h\nu} \mathrm{EtOOC{-}N} \quad 50\% \qquad (5.38)$$

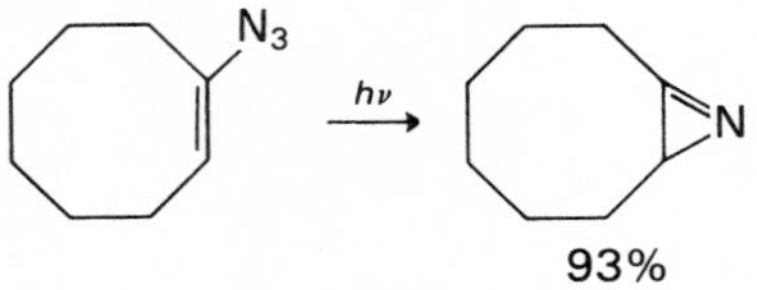

$$\mathrm{N_3} \xrightarrow{h\nu} \mathrm{N} \quad 93\% \qquad (5.39)$$

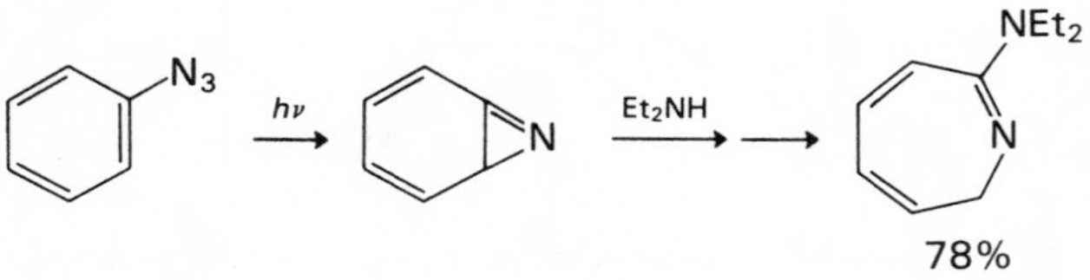

$$\mathrm{N_3} \xrightarrow{h\nu} \mathrm{N} \xrightarrow{\mathrm{Et_2NH}} \longrightarrow \mathrm{NEt_2},\ \mathrm{N} \quad 78\% \qquad (5.40)$$

The photochemistry of diazonium salts ($Ar\overset{+}{N}_2\ X^-$) is important because these compounds have for a long time been widely used for monochromatic photoimaging. Material impregnated with a diazonium salt is exposed to light (usually ultraviolet) through a 'mask' of the object to be recorded. In the exposed areas the diazonium salt is destroyed, and development of the image using a coupler (such as a phenolate salt) that reacts with unchanged diazonium salt to produce an azo-dye provides a direct, positive record of the original. The resolution is very high, the materials are not expensive, and development is simple, so this long-standing process is still used for reproduction of such items as detailed technical plans ('blueprints'). There are many variations in the method of development, but all of these 'diazo' imaging systems rely on the photochemical breakdown of diazonium salts (or their covalent equivalents) by loss of nitrogen. In some hydroxylic solvents, irradiation of a diazonium salt can lead to a phenol in high yield (5.41), and this suggests that bond cleavage to give an aryl cation is a likely mechanism. However, replacement of

$-N_2^+$ by hydrogen may also occur (5.42), and this implies an aryl radical as intermediate. In many systems it is possible that both cationic and radical mechanisms operate.

N_2^+ Cl^- Br Br Br $\xrightarrow[H_2O]{h\nu}$ OH Br Br Br 100% (5.41)

$N_2^+Cl^-$ NO_2 $\xrightarrow[aq.\ HCl]{h\nu}$ + NO_2 → X NO_2 >95% (X = Cl, OH)

$\xrightarrow[EtOH]{h\nu}$ • NO_2 → 77% (5.42)

Nitrites and nitro-compounds

Organic nitrites (R–O–N=O) absorb in the near-ultraviolet region (λ_{max}~360 nm, ϵ_{max} <100 l mol^{-1} cm^{-1}) as a result of a relatively low-energy $n \rightarrow \pi^*$ transition. The exclusive reaction of the electronically excited state so obtained is cleavage of the oxygen–nitrogen single bond, to give an alkoxy radical (RO·) and the nitric oxide (NO). These are responsible for further reaction and product formation; for example, irradiation of *t*-butyl nitrite leads to acetone and nitrosomethane (as its dimer) because the tertiary alkoxy radical fragments to a ketone and an alkyl radical (5.43).

O–NO $\xrightarrow{h\nu}$ NO + O• → O + $\dot{C}H_3$ $\xrightarrow{NO}$ CH_3NO ↓ $(CH_3NO)_2$ (5.43)

Nitrites with a longer alkyl chain give rise to 4-nitrosoalcohols (5.44) as a result of intramolecular hydrogen abstraction by the alkoxy radical; the carbon-centred radical formed in this step combines with nitric oxide, and the nitroso-compound may be isolated (usually as a dimer) or may tautomerize to an oxime. In unconstrained systems the preferred position of hydrogen abstraction is the one that goes through a six-membered, rather than a different size, cyclic transition state. The overall reaction is sometimes called the Barton reaction, and it has been widely used in the functionalization of steroid systems, especially at the angular methyl groups. The transformation of the 11β-nitrite ester of corticosterone 21-acetate to the oxime of aldosterone 21-acetate (5.45) is an important example of such a steroid reaction, providing a reasonable synthetic route to the steroid hormone aldosterone, which is responsible for controlling sodium ion balance in the body and regulating kidney functions. The position of attack within the alkoxy radical is governed by the preferred conformation of the ring systems, and an example of an 11β-nitrite without the unsaturated ketone group in the ring structure reacts at the alternative angular methyl group (5.46).

ONO $\xrightarrow{h\nu}$ O• + NO

OH $\xrightarrow{NO}$ OH, NO

45%

OH, NOH

(5.44)

ONO, O, OAc, O $\xrightarrow{h\nu}$ HON, O, HO, OAc, O

21%

(5.45)

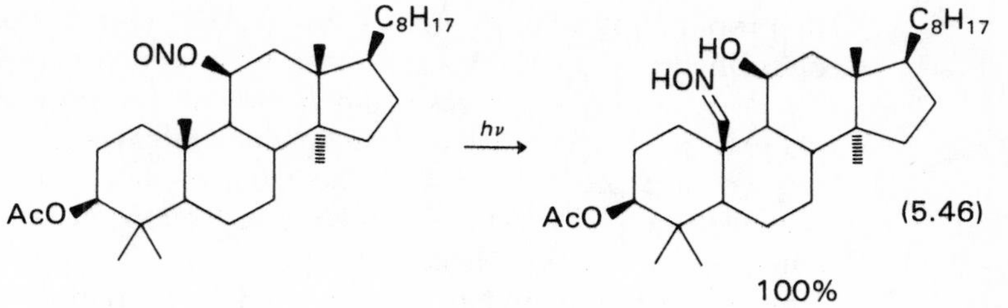

(5.46)

100%

Nitro-compounds (RNO_2) are isomeric with nitrites, but their electronic structure, excited states and photochemistry are very different. There is no very low-lying (n,π^*) state, and nitroalkanes show $n \rightarrow \pi^*$ absorption with a maximum around 275 nm ($\epsilon \sim 20$ l mol $^{-1}$ cm $^{-1}$). In cyclohexane solution, nitromethane (CH_3NO_2) is photoreduced to nitrosomethane (CH_3NO), but nitroethane under the same conditions gives rise to a nitroso-dimer derived from the solvent (5.47). The latter process is probably initiated by cleavage of the carbon–nitrogen bond in the nitroalkane. In basic solution (when the nitroalkane is converted to a nitronate anion) irradiation can lead to efficient formation of a hydroxamic acid (5.48), and this reaction most likely proceeds through formation of an intermediate three-membered cyclic species.

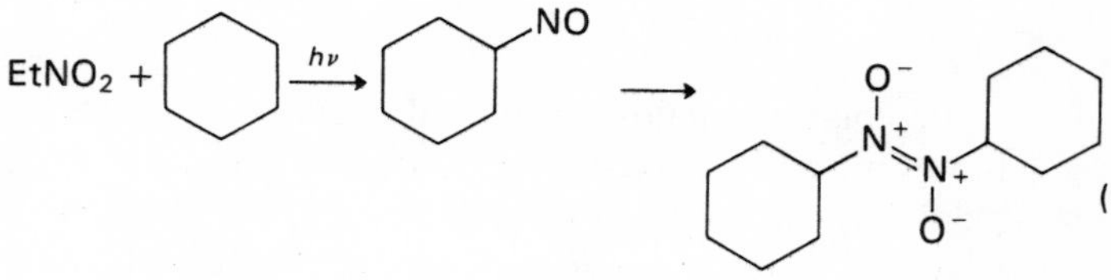

(5.47)

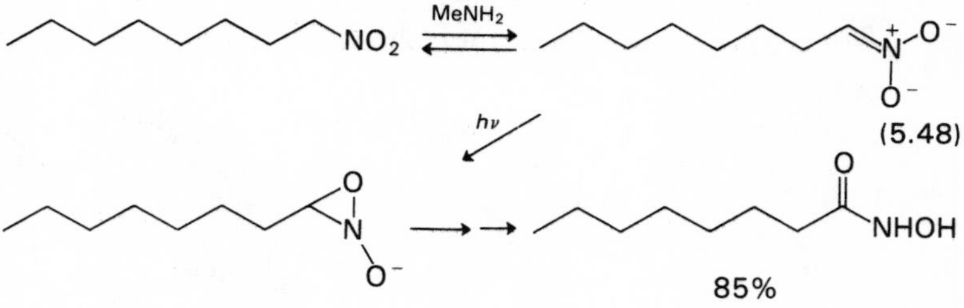

(5.48)

85%

Considerable interest has been shown in the photochemistry of aromatic nitro-compounds, especially those reactions that involve hydrogen transfer to the nitro-group. Nitrobenzene is photoreduced by propan-2-ol, and the major product under some conditions is phenylhydroxylamine (5.49). Other nitro-aromatics behave in a similar way, but in all cases the nature of the products isolated depends on the acidity of the reaction medium and on the details of the work-up procedure, since there is a quite complex array of potential thermal reactions involving reduction products such as nitroso-compounds, hydroxylamines, azoxy-compounds or amines. *o*-Substituted nitrobenzenes can take part in intramolecular hydrogen-transfer reactions, and this accounts for the efficient conversion of *o*-nitrobenzaldehyde to *o*-nitrosobenzoic acid (5.50); it is worth noting that a similar overall reaction observed for *p*-nitrobenzaldehyde in the presence of water takes place by a different mechanism.

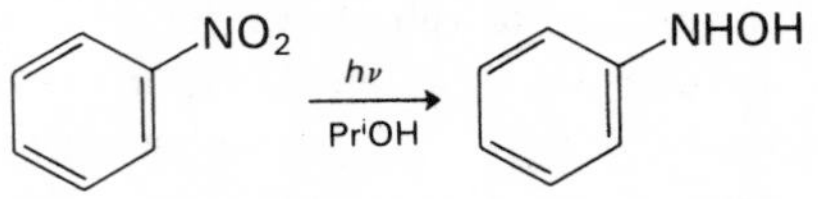

(5.49)

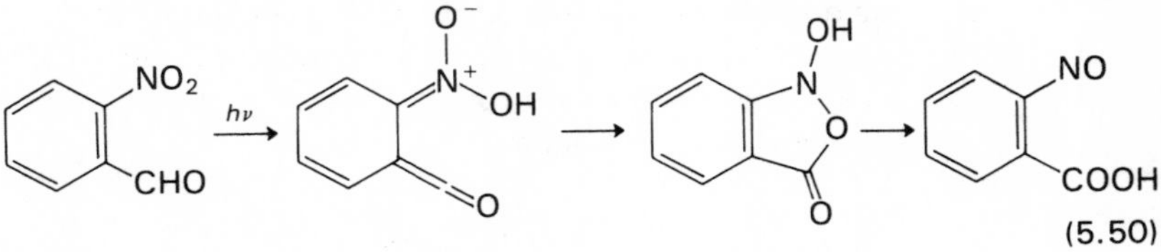

(5.50)

The related reaction for *o*-nitrobenzyl esters results in hydrolysis of the ester, and this has been developed for use in the protection of alcohols, carboxylic acids and amines. For example, the C-1 hydroxyl group in carbohydrates can be protected as its *o*-nitrobenzyl ester, and the ester group removed under the very mild conditions of irradiation in neutral solution (5.51). Similarly, the carboxylic acid

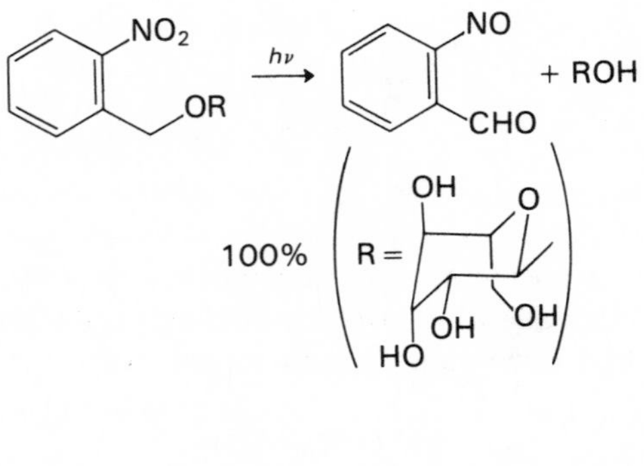

(5.51)

group of an amino acid or peptide may be protected using a related alcohol, and the protecting group removed quantitatively under non-hydrolytic conditions (5.52).

With alkenes, nitrobenzene undergoes (3 + 2) photocycloaddition (5.53), and the 1,3,2-dioxazole derivative so produced can be isolated by working at low temperatures.

NO_2 ; O_2N ; O ; R ; O ; $h\nu$; HO ; R ; O (5.52)

100% (R = CH—NH—COOBu^t ; CH_2Ph)

O ; N^+ ; O^- ; + ; $h\nu$; O ; N ; O (5.53)

Saturated oxygen and sulfur compounds

The electronic structure of saturated compounds containing a heteroatom such as oxygen or sulfur can be described in terms of bonding (σ), antibonding (σ*) and non-bonding (*n*) orbitals. The lowest-energy excited states are probably of the (*n*,σ*) type, although there is evidence for fairly low-lying Rydberg states (see p. 12) in some cases. On this basis it is understandable that bond homolysis is a common outcome of the irradiation of such compounds, because (*n*,σ*) states are repulsive or only weakly bonding, and the energy of the states is generally much in excess of the single-bond dissociation energies within the molecule.

Alcohols and ethers absorb only in the short-wavelength ultraviolet region (< 200 nm). This makes them very useful solvents for many photochemical reactions on account of their transparency to longer-wavelength radiation, but they are reactive towards the excited states of some classes of compound such as ketones. Where such reactivity does compete with the desired processes, the choice of *t*-butanol, which does not contain a readily abstracted alpha-hydrogen atom, may overcome the problem. When they are subjected to high-energy irradiation, alcohols and ethers normally fragment to give eventually both higher and lower molecular weight products. This can be of considerable theoretical interest, but the synthetic potential is severely limited.

Exceptionally, oxiranes (epoxides), which are three-membered cyclic ethers, absorb at somewhat longer wavelengths and cleave more specifically. For 2,3-diaryloxiranes the excited state undergoes heterolytic carbon–carbon bond cleavage, and the coloured dipolar product (5.54) can be studied spectroscopically at low temperatures. In the presence of an electron-transfer sensitizer, diaryloxiranes undergo cycloaddition with a range of multiply bonded addends, leading to high yields of five-membered heterocyclic compounds (compare a similar nitrogen system in equation 5.5); even oxygen can be employed as addend (5.55) if the aryl group is electron-rich.

O, Ar, Ar $\xrightarrow{h\nu}$ Ar, $\overset{+}{O}$, Ar (5.54)

O, Ar, Ar $\xrightarrow[C_{14}H_8(CN)_2]{h\nu,\ O_2}$ Ar, O, Ar, O—O (5.55)

(Ar = *p*-$MeOC_6H_4$) 100%

Dialkyl peroxides (RO–OR), unlike ethers, absorb longer-wavelength ultraviolet radiation (up to about 350 nm); this is caused by an interaction between nominally non-bonding orbitals on the adjacent oxygen atoms, rather like the interaction that makes azo-compounds coloured (see p. 148). Irradiation of peroxides generally leads to oxygen–oxygen bond cleavage, and if no other compound is present the alkoxy radicals may disproportionate or fragment (5.56). More often, a peroxide is employed as a useful source of alkoxy radicals for further reaction such as initiation of vinyl polymerization, or oxidative dimerization (5.57) that involves hydrogen atom abstraction. Dialkyl peroxides are thermally, as well as photochemically, labile, and in many contexts diacyl peroxides, especially dibenzoyl peroxide (5.58), are used as convenient sources of radical initiators because of their stronger ultraviolet absorption and lower thermal reactivity.

Sulfur compounds absorb at longer wavelengths than their oxygen counterparts, and with increased molar absorption coefficients. Dialkyl sulfides (R_2S) cleave on irradiation to generate alkyl (R·) and alkanethiyl (RS·) radicals. In the presence of an added trivalent phosphorus compound such as triethyl phosphite, the sulfur can be abstracted from the thiyl radical, and this offers a way of carrying out desulfuration of a sulfide to give an alkane (5.59). Such procedures have been used to prepare compounds known as cyclophanes (e.g. 5.60), which are of interest for studying the interaction between electrons in aromatic rings that are held in particular orientations in fairly rigid frameworks. The sulfides can be prepared readily from haloalkane and thiol precursors, and the new carbon–carbon bonds are formed on irradiation. A related route to cyclophanes utilizes sulfones rather than sulfides (5.61); one advantage in this case is that extrusion of sulfur dioxide does not require an added scavenger.

$$\text{O}^{\bullet} \quad + \; \dot{C}H_3 \longrightarrow C_2H_6 \qquad (5.56)$$

$$(Bu^tO)_2 + \xrightarrow{h\nu} \qquad (5.57)$$

$$\xrightarrow{h\nu} Ph\dot{C}O_2 \longrightarrow Ph^{\bullet} + CO_2 \qquad (5.58)$$

$$PhCH_2SCH_2Ph \xrightarrow[P(OEt)_3]{h\nu} PhCH_2CH_2Ph \qquad (5.59)$$

$$\xrightarrow[P(OEt)_3]{h\nu} \qquad (5.60)$$

85%

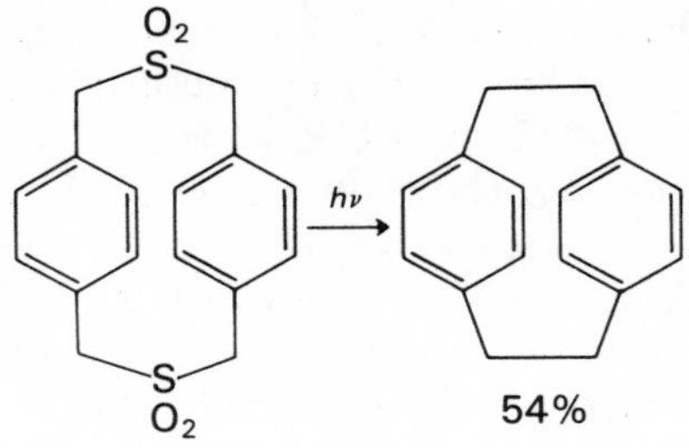

(5.61)

54%

Thiols (RSH) cleave at the sulfur–hydrogen bond on irradiation, and this allows the preparation of sulfides by subsequent radical addition to an alkene (5.62). Hydrogen sulfide behaves in the same way (5.63), and long-chain thiols are made commercially for use as modifiers in emulsion polymerization.

Disulfides (RS–SR) are not as useful as peroxides in their photochemical reactions, partly because carbon–sulfur bond cleavage (to give R. and $RS._2$) competes more effectively with sulfur–sulfur bond cleavage than does C–O with O–O cleavage for peroxides, but largely because thiyl radicals are less reactive than their oxygen analogues.

(5.62)

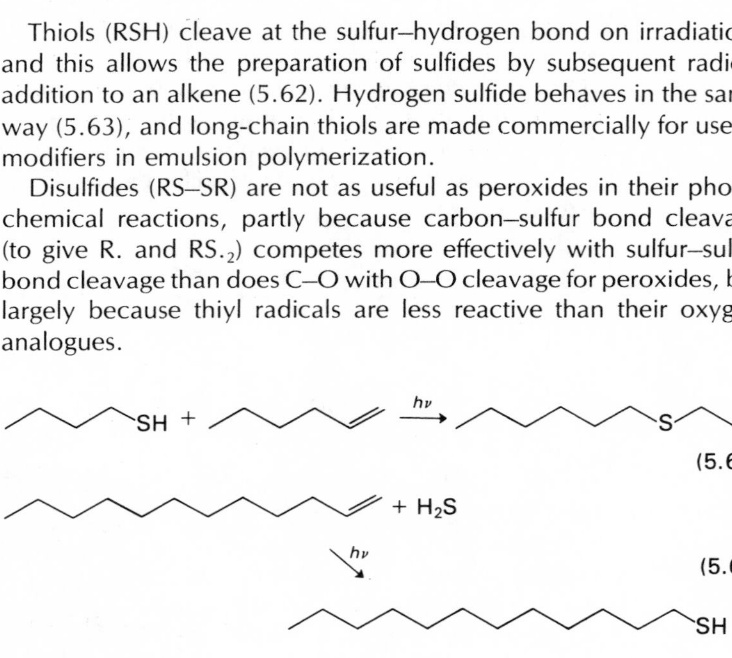

$+ H_2S$

(5.63)

Halogen compounds

Whereas simple haloalkanes play a very important role in synthetic and mechanistic organic chemistry in the ground state, the interest in excited-state reactions centres more on polyhaloalkanes and on vinyl or aryl halides. Chloromethane absorbs only in the far ultraviolet (λ_{max} 173 nm), but longer-wavelength absorption occurs for bromo- or iodo-compounds (CH_3I has λ_{max} 258 nm) and for polyhalo-compounds (iodoform, CHI_3, has λ_{max} 349 nm and is coloured yellow).

The photochemical addition of polyhalomethanes to alkenes has been a known reaction for a long time. Efficient cleavage occurs preferentially at the weakest carbon–halogen bond, and the formation of the adduct (5.64) may well be a chain process.

$$+ \ CBrCl_3 \xrightarrow{h\nu} \text{Br} \ \ CCl_3 \quad 88\% \qquad (5.64)$$

Apart from lengthening the chain by one carbon atom, this reaction provides a compound with two readily modified functional groups. Many other halogen compounds offer a more complex picture. Radical-derived products are sometimes obtained, but in other cases a heterolytic mechanism must operate, for example to give a methyl ether when naphthylmethyl iodides are irradiated in methanol (5.65).

$$\text{Br} \xrightarrow[\text{MeOH}]{h\nu} \text{OMe} \quad 88\% \qquad (5.65)$$

Recent studies on simple haloalkanes, especially iodoalkanes, suggest that the initial process is always homolytic carbon–halogen bond cleavage, but that iodo-systems are especially susceptible to subsequent electron transfer from the alkyl radical to the fairly unreactive iodine atom. This gives the alkyl carbonium ion that reacts with, for example, a nucleophilic solvent. The operation of this mechanism provides for the generation and reaction of vinyl carbonium ions from vinyl iodides (5.66), and this offers one of the few ways of generating such intermediates.

$$\text{I} \xrightarrow{h\nu} \ \bullet + I^{\bullet} \longrightarrow \ ^{+} + I^{-} \xrightarrow{\text{MeOH}} \text{OMe} \quad 65\% \qquad (5.66)$$

With alkenes, di-iodomethane (5.67) or tri-iodomethane (5.68) reacts photochemically to form cyclopropanes. The intermediate is not a carbene, but most likely a carbonium ion ($\overset{+}{C}H_2I$ or $\overset{+}{C}HI_2$), and this makes the reaction a useful complement to the thermal carbene or carbenoid routes to cyclopropanes. This photochemical method works well for electron-deficient alkenes, and also for sterically hindered alkenes.

COOMe + CH_2I_2 $\xrightarrow{h\nu}$ COOMe 71% (5.67)

+ CHI_3 $\xrightarrow{h\nu}$ I 65% (5.68)

Hypohalites (RO–Hal) are similar to nitrates (see p. 155) in their photochemical behaviour. Ultraviolet irradiation gives an (n,π^*) excited state that cleaves to form an alkoxy radical and a halogen atom. The radical may undergo alpha-cleavage before recombination with the halogen atom occurs, and this accounts for the formation of 5-iodopentanal (5.69) from the hypoiodite of cyclopentanol; such hypoiodites are generated *in situ* from the alcohol, iodine and mercury(II) oxide. In open-chain systems the alkoxy radical can

OI $\xrightarrow{h\nu}$ O• + I• ⟶ O ; I· ⟶ O I (5.69)

OCl $\xrightarrow{h\nu}$ O• + Cl• ⟶ OH $\xrightarrow{Cl\cdot}$ Cl OH 80% (5.70)

abstract a hydrogen atom from within the molecule, and the product is a 4-haloalcohol (5.70), in which an 'unactivated' position in the original alkyl group now carries a reactive functional group. As with nitrites, this reaction has found applications in steroid and terpene chemistry.

N-Chloroamines (R_2NCl) react in a similar way, although strongly acidic conditions are required for efficient reaction. This suggests that the mechanism involves protonated species, and a typical example is the formation of an *N*-substituted pyrrolidine from an *N*-chlorobutylamine (5.71) by irradiation in acid followed by alkaline work-up. Pyrrolidin-2-ones are formed under non-acidic conditions from *N*-haloamides (5.72). The related cleavage of *N*-chlorosuccinimide has been described previously as a β-cleavage reaction of an excited carbonyl compound (see p. 115), and *N*-haloamides can undergo (internal) photoaddition to alkenes in an exactly analogous way (5.73).

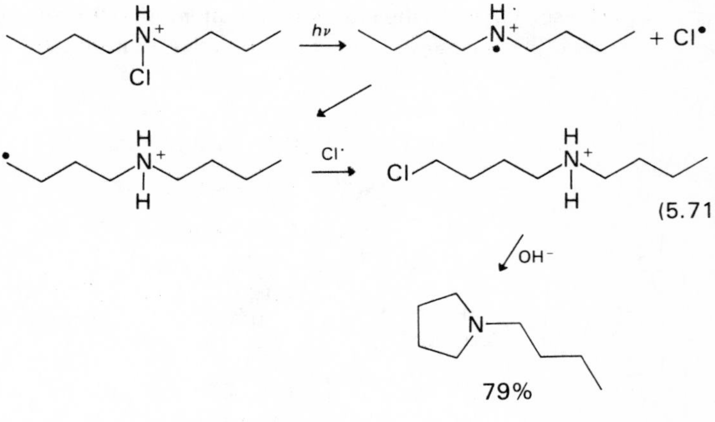

(5.71)

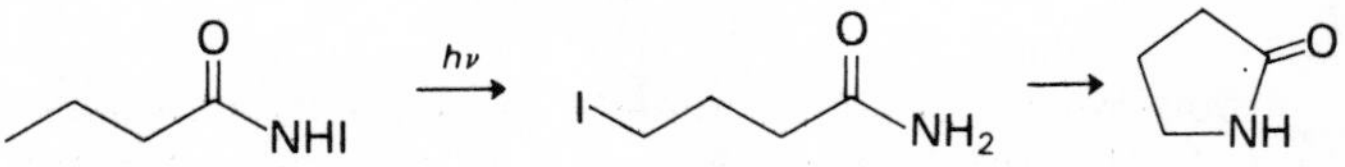

79%

(5.72)

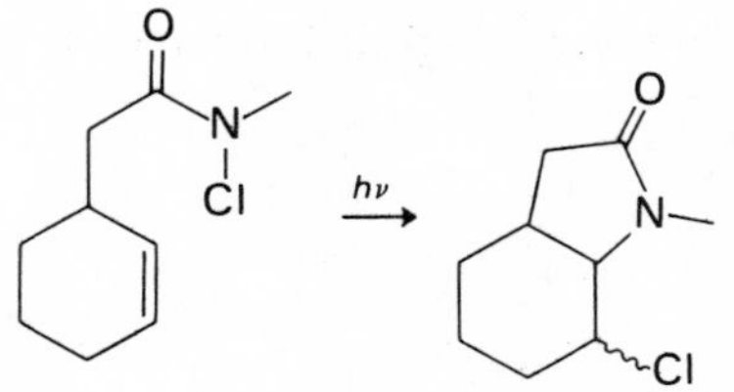

(5.73)

Photohalogenation and photonitrosation

Several commercially important photochemical processes involve the reaction of a halogenated compound with an organic substrate, and although it is the non-organic component that is excited, it is appropriate to describe them in this section.

The reaction between chlorine or bromine and an alkane is a chain reaction that can be initiated photochemically (5.74) or thermally; quantum yields in the light-promoted process can be extremely high (more than 10^6). Photochemical initiation offers advantages in enabling the operating temperature to be kept low, and in making for easier controls on a large manufacturing scale. Alkanes often react fairly unselectively, especially with chlorine, to give a mixture of isomeric products (replacement of non-equivalent hydrogen atoms) and products with different degrees of halogenation (replacement of one, two or more hydrogen atoms). For some purposes, such as use as solvents, mixtures may be acceptable, but for others, such as for intermediates in organic synthesis, further purification is necessary. The selectivity of the reaction is greater when bromine is used (5.75), since the bromine atom is less reactive than the chlorine atom.

66%

$$Cl_2 \xrightarrow{h\nu} 2Cl^\bullet$$

$$Cl^\bullet + RH \longrightarrow HCl + R^\bullet \qquad (5.74)$$

$$R^\bullet + Cl_2 \longrightarrow RCl + Cl^\bullet \quad \text{etc.}$$

$$(CH_3)_3CH \xrightarrow[Cl_2]{h\nu} \underset{40\%}{(CH_3)_3CCl} + \underset{60\%}{(CH_3)_2CH{-}CH_2Cl}$$

$$(CH_3)_3CH \xrightarrow[Br_2]{h\nu} \underset{100\%}{(CH_3)_3CBr} \qquad (5.75)$$

Substrates that carry a replaceable benzylic hydrogen atom, or a similar hydrogen that gives rise to a stabilized radical, can be selectively chlorinated or brominated. Ethylbenzene leads to only

$$C_6H_5CH_2CH_3 + Br_2 \xrightarrow{h\nu} C_6H_5CH(Br)CH_3 \qquad (5.76)$$

one mono-brominated derivative (5.76), and with five molar equivalents of chlorine, *o*-xylene gives a very high yield of the pentachloro product (5.77), with very little contamination by hexachloro and tetrachloro compounds.

$$\text{o-xylene} + 5Cl_2 \xrightarrow{h\nu} C_6H_4(CCl_3)(CHCl_2)\ \ 98\% \qquad (5.77)$$

Aromatic C–H bonds are not broken in radical halogenation, because they are a little stronger than aliphatic C–H bonds. When benzene reacts photochemically with chlorine, a radical *addition* process takes place, and the mixture of stereoisomeric hexachlorocyclohexanes (5.78) includes one isomer which has powerful insecticidal properties but which, unlike some chlorinated insecticides, is readily biodegradable.

$$C_6H_6 + 3Cl_2 \xrightarrow{h\nu} C_6H_6Cl_6\ \ 95\% \qquad (5.78)$$

The halogenation of alkanes in the presence of sulphur dioxide yields alkanesulphonyl chlorides (5.79), and these are made in large quantities for conversion to metal alkanesulphonates (used as emulsifiers in polymerizations) or to nitrogen-containing derivatives. The sulphur dioxide acts by trapping the alkyl radical; it does not terminate the chain mechanism, and so quantum yields can be very high (~2000).

$$\begin{aligned} Cl_2 &\xrightarrow{h\nu} 2Cl^\bullet \\ Cl^\bullet + RH &\longrightarrow HCl + R^\bullet \\ R^\bullet + SO_2 &\longrightarrow RSO_2^\bullet \\ RSO_2^\bullet + Cl_2 &\longrightarrow RSO_2Cl + Cl^\bullet \text{ etc.} \end{aligned} \qquad (5.79)$$

Photonitrosation is related to photohalogenation; the absorbing species is nitrosyl chloride (NOCl), the reactive intermediate is a chlorine atom formed by homolytic bond cleavage, and the product (5.80) is an oxime that arises by tautomerization of an initial nitrosoalkane. The reaction does not go by way of a chain mechanism, so quantum yields are not high, but one company has found it economically viable to operate on a very large scale for the production of cyclohexanone oxime from cyclohexane (5.81).

$$\begin{aligned} Cl{-}N{=}O &\xrightarrow{h\nu} Cl^{\bullet} + NO \\ Cl^{\bullet} + R_2CH_2 &\longrightarrow HCl + R_2\dot{C}H \\ R_2\dot{C}H + NO &\longrightarrow R_2CH{-}NO \\ R_2CH{-}NO &\longrightarrow R_2C{=}NOH \end{aligned} \qquad (5.80)$$

$$\text{cyclohexane} + NOCl \xrightarrow{h\nu} \text{cyclohexanone oxime } (>80\%) \longrightarrow \text{caprolactam} \longrightarrow \text{nylon-6} \qquad (5.81)$$

This oxime rearranges under acidic conditions to a seven-membered cyclic amide (caprolactam), that is the monomer used in making one of the most widely used types of nylon, namely nylon-6. A more specialized polyamide, nylon-12, is made in a similar way from cyclododecane (5.82); this polymer finds applications in coating metals and in making plastic components for motor cars.

$$\text{cyclododecane} + NOCl \xrightarrow{h\nu} \text{cyclododecanone oxime } (>70\%) \qquad (5.82)$$

Photo-oxidation of alkanes

Alkanes absorb only in the far ultraviolet region (ethane has λ_{max} 135 nm, $\epsilon_{max} \sim 10^4$ l mol^{-1} cm^{-1}, corresponding to a $\sigma \rightarrow \sigma^*$ transition), and their high-energy photochemistry is largely a matter of fragmentation as a result of C–H and/or C–C bond cleavage. However, irradiation at slightly longer wavelengths in the presence of oxygen gives a hydroperoxide (5.83); the absorbing species is either oxygen itself or a weak charge-transfer complex between oxygen and the hydrocarbon. Under conditions of electron-transfer sensitization, using 9,10-dicyanoanthracene as sensitizer, strained aryl-substituted cyclopropanes (5.84) or cyclobutanes are converted into cyclic peroxides. The excited sensitizer takes an electron from the hydrocarbon, to leave a radical cation that is capable of reaction with ground-state oxygen.

$$+ O_2 \xrightarrow[(185\ nm)]{h\nu} \text{OOH} \qquad (5.83)$$

$$\text{Ph, Ph} \xrightarrow[C_{14}H_8(CN)_2]{h\nu,\ O_2} \text{Ph, Ph, O—O} \qquad (5.84)$$

Poly(alkenes) such as polythene are almost entirely alkane in nature, but many such polymers are degraded by exposure to light under normal conditions; they lose mechanical strength and many become brittle or powdery. Ultrapure polythene in the absence of air is photochemically stable, and it seems that photo-oxidation promoted initially by impurities introduced during processing is responsible for the normal photodegradation. The impurities may well be hydroperoxides, and a mechanism for chain cleavage (5.85) provides a carbonyl group that can absorb light, abstract a hydrogen

OOH $\xrightarrow{h\nu}$ $O^{\bullet}$ + $HO^{\bullet}$

O

$h\nu$

O_2

OOH

(5.85)

atom from elsewhere in the polymer chain, and so open up the way for further hydroperoxide formation by reaction of polymer chain radicals with oxygen.

Stabilization of polymers against such photo-oxidation can be achieved by incorporating a photochemically stable ultraviolet absorber (e.g. carbon black, or 2-hydroxybenzophenone, 5.86) that prevents absorption by the hydroperoxide, or a radical scavenger (e.g. *p*-dihydroxybenzene, 5.87), or a compound that catalyses the decomposition of hydroperoxides (e.g. metal dithiocarbamates, 5.88). In some contexts accelerated photodegradation may be desirable, such as for agricultural use as a mulching film, or (less practicably) to avoid litter problems arising from the long life of normal plastics. The use of inorganic sensitizers for the photo-oxidation can help achieve this.

(5.86)

(5.87)

$R_2N-C(S)_2Ni(S)_2C-NR_2$ (5.88)

Further reading

A. Padwa, *Chemical Reviews*, vol. 77, American Chemical Society (1977), p. 37. Although there are more recent reviews of specialized aspects of C=N photochemistry, this a survey of all the photoreactions.

P. S. Mariano, in W. M. Horspool (ed.), *Synthetic Organic Photochemistry*, Plenum (1984). This is a comprehensive account of the photochemical reactions of iminium salts and related species.

P. S. Engel, *Chemical Reviews*, vol. 80, American Chemical Society (1980), p. 99. The thermal and photochemical reactions of azoalkanes are covered in this review.

R. W. Binkley and T. W. Flechtner, in W. M. Horspool (ed.),

Synthetic Organic Photochemistry, Plenum (1984). The use of nitro groups features prominently in this article about photoremovable protecting groups in organic synthesis.

E. Brinckman, G. Delzenne, A. Poot and J. Willems, *Unconventional Imaging Systems*, Focal Press (1978). This short book provides a good account of diazo imaging systems, set in the context of other non-silver processes.

Y. L. Chow, in S. Patai (ed.), *The Chemistry of Amino, Nitroso, Nitro Compounds*, Wiley (1982), p. 181. The photochemistry of nitroso and nitro compounds is extensively detailed.

P. J. Kropp, *Accounts of Chemical Research*, vol. 17, American Chemical Society (1984), p. 131. This account of the photochemistry of haloalkanes in solution presents evidence for the intermediacy of radical, cationic and carbene species.

J. D. Coyle, R. R. Hill and D. R. Roberts (eds), *Light, Chemical Change and Life*, Open University Press (1982). In this book are chapters on the photodegradation of polymers, photohalogenation, and the large-scale manufacture of caprolactam.

Index

ABIN, 149
absorption coefficient, 13
absorption of light, 2, 9
absorption spectra, 13
acetal formation, 119
acetone, addition reactions, 116
  cycloaddition reactions, 127
  decarbonylation, 108
  excited state energies, 17, 29
  lifetimes, 19
acetophenone, photoreduction, 116
acetophenone oxime, 142
acetylene, (π,π^* state), 15
acetylenes, 72
acetylnaphthalene, absorption
  spectrum, 18
acridine *N*-oxides, 104
acrylonitrile, 61, 92, 96, 139
actinometer, 33, 100
acyl halides, 110
addition, of methanol, 116
addition reactions, alkenes, 58
  alkynes, 73
  aromatic compounds, 90
alcohols, 159
aldehydes, 107, 110
aldosterone, 156
alkanes, 166
  photo-oxidation, 168
alkanesulphonates, 167
alkene, energy diagram, 41
alkenes, 40
  addition reactions, 58
  cycloaddition reactions, 61
  electronic configuration, 12
  isomerization, 56
  photo-oxidation, 69
alkynes, 72
  cycloaddition reactions, 93, 95, 129
allenes, cycloaddition reactions, 93
amines, diaryl, 101
anisole, 81, 92
antarafacial, 52
anthracene, 95
  fluorescence spectrum, 21
anthraquinone dyes, 118
antibonding orbital, 10
aromatic compounds, addition
  reactions, 90
  cyclization reactions, 97
  cycloaddition reactions, 91
  ring isomerization, 86
  substitution reactions, 77
aryl halides, 82
azides, 151
aziridines, 144
2*H*-azirines, 143
azobenzene, 148, 150
azo-compounds, 148
azomethane, 149

Barton reaction, 156
Beer–Lambert law, 13
benzaldehyde, 110, 130
  o-nitro, 158
benzene, additional reactions, 90
  cycloaddition reactions, 91
  excited state energies, 17
  lifetimes, 19
  ring isomerization, 88

benzoate esters, 128
benzoins, 109
benzonitrile, 92, 147
benzophenone, as triplet sensitizer, 30
  cycloaddition reactions, 126
  excited state energies, 17
  lifetimes, 19
  photoreduction, 116
benzophenones, *o*-alkyl, 125
benzoquinone, cycloaddition reactions, 127
benzvalene, 86
bicyclobutane, 56
biological damage, 65
biphenyl, 2-azido, 153
biphenyl lifetimes, 19
blueprints, 154
bonding orbital, 9
β-bourbonene, 65
bromobenzene, 82
buta-1,3-diene, 47, 56, 63
  excited state energies, 17
but-2-ene, 62

caprolactam, 168
  N-phenyl-, 85
carbenes, 151
carbonyl compounds, 106
  cycloaddition reactions, 126
  hydrogen abstraction, 115
  rearrangements, 131
carboxylic acids, 107, 110
chemiluminescence, 3
chloroacetamides, 102
N-chloroamines, 165
chrysenes, 99
CIDNP, 36
cis–trans isomerization, 42, 124, 148
α-cleavage, 107
concerted reactions, 5
conformational effects, 50, 53, 56
conrotatory, 48
cyclization reactions, aromatic compounds, 97
cycloaddition reactions, alkenes, 61
  alkynes, 73
  aromatic compounds, 91
  azo-compounds, 151
  carbonyl compounds, 126
  imines, 145
  nitriles, 148
  nitrobenzene, 159
  thioketones, 137
cyclobutanols, 120
cyclobutanones, 114
cyclobutene, 47
cyclodecanone, 123
cyclododecane, 168
cyclododecanone, 121
cycloheptatriene, 50
  1-cyano, 53
cycloheptene, 68
  1-methyl, 60
cyclohexa-1,3-diene, 47
cyclohexadienones, 131
cyclohexane, 168
cyclohexanone, 112, 116
cyclohexene, 68
  1,2-dimethyl, 57, 60
  1-iodo, 163
  1-methyl, 60
  1-phenyl, 59
cyclohex-2-enone, 65
cyclohexenones, 131
cyclo-octa-1,3-diene, 43
cyclo-octa-1,3,5-triene, 47
cyclo-octene, 58, 70
  1-phenyl, 45
cyclo-oct-2-enone, 124
cyclo-oct-3-enone, 111
cyclopentene, 63
cyclopent-2-enone, 65, 74

decarbonylation, 108
decarboxylation, 109
degradation of polymers, 69, 169
delocalized orbital, 10
Dewar benzene, 87
diazirines, 152
diazo compounds, 151
diazo imaging systems, 154
diazonium salts, 151
dibenzodioxins, 84
dibenzoyl peroxide, 151
dichloroethylene, *cis–trans* isomerization, 42
α-diketones, 121
diketopiperazine, 161

dimerization, anthracene, 96
di-π-methane reaction, 54, 73, 132
dioxetanes, 71
dipole-dipole quenching, 28
dipole moment, 15
disrotatory, 48
disulfides, 162
1,4-dithianes, 137
dithietanes, 137
DNA, 65
dyes, light-fastness, 148

electrocyclic process, 46
electrocyclic reactions, stereochemistry, 48
electromagnetic radiation, 10
electron acceptors, 6
electron distribution, 5
electron donors, 6
electron transfer, 7, 27, 58
energy transfer, 27
epoxides, 160
ethers, 159
  diaryl, 101
ethylbenzene, 166
  absorption spectrum, 18
ethylene, 1,1-diphenyl, 58
  (π,π*) state, 15
exchange quenching, 28
excimer, 27
exciplex, 27, 128
excited state, 2
  classification, 12
  dynamic properties, 20
  energy, 17, 21
  lifetime, 19, 36
  non-vertical, 42
  redox properties, 7
  static properties, 15
  vertical, 42

filter solution, 38
flash photolysis, 35
fluorescence, 20
  exciplex, 27
  spectra, 17
fluoroanisole, 78
formaldehyde, dipole moment, 16
  (*n*,π*) state, 15
formamide, 110
Franck–Condon principle, 15
fulvene, 88
furan, cycloaddition reactions, 96
furans, 89

geometrical isomerization, 42
Grotthuss–Draper Law, 3
ground state, 2

haloalkanes, 162
N-haloamides, 165
hexachlorocyclohexane, 90, 167
hexa-1,3-diene, 51
hexa-2,4-diene, 49
hexa-1,3,5-triene, 47, 69
hydrogen abstraction, carbonyl compounds, 115
hydroperoxides, 168
hypohalites, 164

imides, addition reactions, 118
  cycloaddition reactions, 130
  cyclization, 123
imines, 142
  cyclization, 101
iminium salts, 142, 146
indene, 71
internal conversion, 23
intersystem crossing, 24, 46
iodoalkanes, 163
iodoform, 163
isomerization, alkenes, 56
  aromatic rings, 86
  geometrical, 42, 124, 142, 148

Jablonski diagram, 25

Kasha's rule, 20
ketones, 5, 106

β-lactams, 112, 152
lifetime, excited state, 19
  fluorescence, 22
  phospherescence, 23
light-fastness, 148
limonene, 60
localized orbital, 10
luminescence, 20

maleic anhydride, 74, 92
maleonitrile, 128
matrix isolation, 37
mercury lamps, 37
meta-cycloaddition, 91
methoxybenzenes, substitution
  reactions, 80
methylene blue, 70
molecular orbital diagram, 2
molecular orbitals, 9
myrcene, 63

naphthalene, cycloaddition reactions,
  151
  Jablonski diagram, 26
  radical cations, 81
  reduction, 91
2-naphthol, 95
nitrenes, 153
nitriles, 142, 147
nitrites, 155
nitroanisole, 78
nitrobenzene, 157, 159
o-nitrobenzyl esters, 158
nitro-compounds, 155
nitromethane, 156
nitrosyl chloride, 167
non-bonding orbital, 10
non-radiative process, 20, 23
norbornadiene, 63
norbornene, 61, 68
Norrish type 1 reaction, 108
Norrish type 2 reaction, 120
nylon, 168

octa-1,3,5,7-tetraene, 47
octa-2,4,6-triene, 48
orbital symmetry, 5
ortho-clycloaddition, 91
oxa-di-π-methane rearrangement, 110
oxetanes, 126
oxetes, 126
N-oxides, 103
oxiranes, 160
oxygen, 30, 69, 97, 168

para-cycloaddition, 91
penta-1,3-diene, excited state energies,
  29
pentan-3-one, 108
perepoxide, 72
pericyclic reactions, 5, 51
peroxides, 160
phenanthrene, 97
phenyl azide, 154
phosphorescence, 22
photochemical mechanisms, 30
photochemical process, 1, 20
photochemistry, preparative, 37
photochromism, 145
photoequilibrium, 8
photo-Fries reaction, 84
photohalogenation, 166
photon, 10
photonitrosation, 166
photo-oxidation, alkanes, 168
  alkenes, 69
  thioketones, 139
photophysical process, 1, 20
photostationary state, 8, 43, 88
phototendering, 118
phthalimides, 123, 130
pi orbital, 10
pigments, light-fastness, 148
polythene, 169
prismane, 88
propanal, 128
provitamin D, 51
pyrene, lifetimes, 19
pyridazines, 86
pyridine *N*-oxide, 104
pyrroles, 89
pyruvate esters, 121

quadricyclane, 63
quantum yield, 22
  measurement, 33
quenching, 27
  rate constant, 34
quinoline *N*-oxide, 105
quinones, photoreduction, 118

radiative process, 20
radical anions, 7, 82, 118
radical cations, 7, 81
radical ions, alkenes, 58, 67
radical reactions, aromatic compounds,
  84

rate constants, photochemical, 33
rearrangements, carbontyl compounds, 131
reduction, azobenzenes, 151
  carbonyl compounds, 115
  imines, 144
  naphthalene, 91
  nitro-compounds, 157
  thioketones, 137
retinal, 44
rhodopsin, 44, 146
ring isomerization, aromatic compounds, 86
Rose Bengal, 70
Rydberg state, 12
Rydberg states, alkenes, 40, 57

salicylaldimines, 145
selection rules, 32
self-quenching, 27
sensitization, 30, 45
sigma orbital, 10
sigmatropic shifts, 51
singlet oxygen, 30, 69, 97, 139
singlet state, 14
  energy, 17
  lifetimes, 19
solvent effects, 32, 133
solvents, for photochemistry, 38, 116, 147, 159
Stark–Einstein law, 4
Stern–Volmer plot, 34
stilbene, absorption spectrum, 13
  cis–trans isomerization, 42
  cyclization, 97
  excited state energies, 17
styrenes, addition reactions, 58
substitution reactions, aromatic compounds, 77
succinimide, *N*-chloro, 115
sulfides, 161
  diaryl, 101
sulfochlorination, 167
superoxide ion, 72
suprafacial, 52, 63

thermodynamic factors, 5
thietanes, 138
thietes, 138
thiobenzophenone, 138
thiocarbonyl compounds, 136
thioketones, 136
thiols, 162
thiophenes, 89
thioxanthone, 139
thymine, 65
triplet energy, determination, 46
triplet quencher, 29
triplet sensitization, 45
triplet state, 14
  energy, 17
  lifetimes, 19

units in photochemistry, 11
α,β-unsaturated ketones, 65, 124

valence isomer, 86
vitamin A, 46
vitamin D, 50

Woodward–Hoffmann rules, 6, 48

xylene, 86
  chlorination, 167